Composite Structures of Steel and Concrete

Volume 1. Beams, Columns, Frames and Applications in Building

The Constrado Monographs deal with the application of
steel in construction. They each treat a specific subject,
and the texts are written with authority and expedition.
Subjects are treated in depth and are taken to the point
of practical application.

Advisory Editor
M. R. Horne, MA, ScD, FICE, FIStructE
Professor of Civil Engineering, University of Manchester

Already published

The Stressed Skin Design of Steel Buildings
by E. R. BRYAN, MSc, PhD, FICE, FIStructE
Professor of Structural Engineering, University of Salford

Thin Plate Design for Transverse Loading
by B. AALAMI, BSc, DIC, PhD, CEng, MICE, MRINA
Associate Professor of Structural Mechanics, Arya Mehr University of
Technology, Tehran, Iran
and D. G. WILLIAMS, BE, MS, DIC, PhD
Project Engineer, Redpath Dorman Long, Bedford

CONSTRADO MONOGRAPHS

Composite Structures of Steel and Concrete
Volume 1. Beams, Columns, Frames and Applications in Building

R. P. JOHNSON, MA, MICE, FIStructE
Professor of Civil Engineering, University of Warwick

A HALSTED PRESS BOOK

JOHN WILEY & SONS — NEW YORK

Published in the U.S.A.
by Halsted Press, a Division
of John Wiley & Sons, Inc.
New York

First published in Great Britain 1975 by Crosby Lockwood Staples

Library of Congress Cataloging in Publication Data

Johnson, Roger Paul.
     Composite structures of steel and concrete.

     (Constrado monographs)
     'A Halsted Press book.'
     Includes bibliographical references and index.
     CONTENTS: v. 1. Beams, columns, frames and applications in building.
     1. Composite construction. I. Title.
TA664.J63 1975   624'.1834   75–16026
ISBN 0–470–44497–5

Printed in Great Britain

Preface

This book is published in the Constrado Monograph series as a result of discussions between Dr J. C. Chapman, the first Director of Constrado, and the author. The extensive experience of both Dr Chapman and Professor Johnson in research on composite structures gave them a common recognition of the desirability of publishing a book which would review current knowledge of the behaviour of composite structures of steel and concrete, particularly in view of the current development in Britain of codes of practice based on limit state design philosophy for steel, concrete and composite structures.

It is of interest that the author first planned a book of this nature in 1970, in the belief that, within a year or two of that date, all the codes would have been published. Four years later, new design procedures are being developed and tried out at a faster rate than before, due to the introduction of contract drafting of codes of practice. There has been no static period, and none can be foreseen. In spite of the increased pace of the work, no limit state codes of practice for steel or composite structures are yet available, and no codes of any kind for composite columns or frames. The reason is that it has proved to be very difficult to draft so many new design documents at once, due partly to the problems raised by the new design philosophy, and partly to the need for design methods for composite structures to be compatible with those for both steel and concrete structures, which have in the past been developed independently of one another.

It is now hoped that draft design methods for composite structures for both buildings and bridges will be issued in Great Britain for public comment during 1975, and this makes the appearance of this book at this time particularly appropriate.

When faced with a new code based on considerable departures from previous design philosophies, the engineer's attention is necessarily directed towards the detailed requirements, and it is only too easy to put off indefinitely the fulfilment of good resolutions that one will bring oneself up-to-date with the relevant theory. This can lead to a too narrow interpretation of the code, to the detriment of sound, economic design, since no code can ever take the place of professional competence and responsibility. This book, and the companion volume to be published in due course, will provide the engineer with just that systematic summary of the theory, its relationship to design procedures, and examples of its application, that should enable him to tackle the design of steel–concrete composite structures with the desirable background of knowledge and understanding.

M. R. Horne
August 1974

Contents

Notation

A cross-sectional area
a distance from crack to surface of nearest reinforcing bar
B breadth of concrete flange
b breadth of member; effective breadth
c carry-over factor; cover
D overall depth of filled tube
d dimension in cross-sectional plane
E Young's modulus of elasticity
e eccentricity of loading
F force
f direct stress
g dead load per unit length or area
h depth of member
I second moment of area of cross-section
K property of composite column; constant of integration
k connector modulus; creep coefficient; constant
L span of beam; length of shear surface; effective length of column
l length of column between joints
M bending moment
m modular ratio; magnification factor
N normal force; axial force; number of shear connectors

P force on shear connector

p pitch or spacing of shear connectors; reinforcement ratio

Q longitudinal shear per unit length on shear plane in concrete

q longitudinal shear per unit length; imposed load per unit length or area

R span–depth ratio

r radius of gyration; degree of shear connection; vertical force per unit length

s slip; unit stress

T tensile force

t thickness of plate

V vertical shear force

v shear stress

w load per unit length or area; width of crack

x dimension along a member; depth of neutral axis

y depth to plastic neutral axis

Z section modulus

z lever arm

α nondimensional coefficient; concrete contribution factor

β nondimensional coefficient; ratio of end moments

γ partial safety factor

δ deflection; small increment

ε direct strain; differential longitudinal strain

η ratio N_u/M_p for a column cross-section

θ rotation

λ nondimensional coefficient; safety factor; load factor

μ nondimensional coefficient; ratio M_p'/M_p

Φ force ratio, $A_r f_{ry}/A_s f_y$

ϕ curvature; diameter of reinforcing bar

$'$ as a superscript, indicates that the symbol relates to a negative-moment region, as in M_p'

Suffixes

a allowable; axial

b bottom

c concrete; collapse; compression; column; centroid; cylinder

cr crack; critical

cu   cube
d   design
e   effective
f   flange; fixed end; flexure; full interaction; (in γ_f) loading
g   centroid; dead load; gross
k   characteristic
m   material; average (mean)
n   normal; negative
p   plastic; propped
q   imposed load
r   steel in reinforcing bars
s   steel; slab; shear
t   top; tension
u   ultimate; unpropped
w   web; works (cube strength); width (of crack)
x x-axis
y y-axis; yield

Foreword

In 1956, the writer was designing a reinforced concrete structure for a heavily loaded laboratory building for a university. It became evident that the insertion of heavy rolled steel sections into the columns in the lower storeys would be economical, if the combined section could be designed as reinforced concrete; but at that time no such design method was available. Nor was much data available on composite beams, where the scope for economy can exceed that in columns.

To provide data on which rational design methods could be based, a programme of research on such structures was commenced in 1959 at the University of Cambridge. The results of this work, now continued at the University of Warwick, provided the major source of material for this book.

In the belief that research in engineering is incomplete until the results are capable of being used in practice, the writer has also taken part in the development of design methods, based on this and other research, for two British codes of practice at present in draft. [15, 23] The explanation and discussion given here of problems that arose in this work should help both students and practising engineers to understand and to design composite structures and also to interpret the relevant codes of practice. The book is not and cannot be a commentary on the two new 'limit state' codes, for they have yet to be issued for comment and are subject to revision before publication.

The present volume is intended to include all material of a fundamental nature that is applicable to structures for both buildings and bridges, and more detailed information and a worked example relating to buildings. Limit state design philosophy and SI units are used throughout. A summary of the scope of Volume 1 is given in Chapter 1. Subjects that are mainly relevant to bridges have been reserved for Volume 2. These include design for repeated loading, composite plate and box girders, two-way composite action in intersecting beams and in composite plates, and the design of columns whose weight is not negligible in comparison with the load carried.

All the design methods discussed are illustrated by sample calculations, and for this purpose a simple problem, or variations of it, has been used throughout the volume. The reader will find that the dimensions, loadings, and strengths of materials for this structure soon remain in the memory. It is hoped that this procedure gives clarity of exposition without surfeit of arithmetic. In criticising the resulting design, it should be remembered that the object here has been to create and discuss design problems, whereas in practice one rightly tries to avoid them.

An attempt has been made to use a consistent notation throughout the volume. It has been based on Appendix F, 'Preparation of Notations', of CP 110: 1972[9] and so differs in some respects from that commonly used for structural steelwork. The only deliberate departure from CP 110 is the use of m, rather than α_e, for modular ratio.

The writer has shared the excitements of research on composite structures with many colleagues, including research students and undergraduates, and since 1968 has shared the frustrations of drafting codes of practice with the members of two committees of the British Standards Institution. The substantial contributions made by these friends and colleagues to the author's understanding of composite structures are gratefully acknowledged. Responsibility for the ideas and design methods here presented rests with the writer, who would be grateful to be informed of any errors that may be found. Finally, a special word of thanks is due to Mr G. Bernard Godfrey for his early interest in the author's work on composite structures, which led to generous support by the members of the British Constructional Steelwork Association for research on this subject over a period of ten years.

R. P. Johnson.
August 1974.

CHAPTER ONE

Introduction

1.1 Composite beams and slabs

The design of structures for buildings and bridges is mainly concerned with the provision and support of load-bearing horizontal surfaces. Except in long-span bridges, these floors or decks are usually made of reinforced concrete, for no other material has a better combination of low cost, high strength, and resistance to corrosion, abrasion, and fire.

The economical span for a reinforced concrete slab is little more than that at which its thickness becomes just sufficient to resist the point loads to which it may be subjected or, in buildings, to provide the sound insulation required. For spans of more than a few metres it is cheaper to support the slab on beams or walls than to thicken it. When the beams are also of concrete, the monolithic nature of the construction makes it possible for a substantial breadth of slab to act as the top flange of the beam that supports it.

At spans of more than about 10 m, and particularly where the susceptibility of steel to damage by fire is not a problem, as for example in bridges and multi-storey car parks, steel beams become cheaper than concrete beams. It used to be customary to design the steelwork to carry the whole weight of the concrete slab and its loading; but by about 1950 the development of shear connectors had made it practicable to connect the slab to the beam, and so to obtain the T-beam action that had long been used in concrete construction. The term 'composite beam' as used in this book refers to this type of structure.

The same term is used for beams in which prestressed and in-situ concrete act together, and there are many other examples of composite action in structures, such as between brick walls and the beams supporting them, or between a steel-framed shed and its cladding; but these are outside the scope of this book.

No income is received from money invested in the construction of a multi-storey building such as a large office block until the building is occupied. For a construction time of two years, this loss of income from capital may be 15% of the total cost of the building; that is, it may be over half the total cost of the structure. The construction time is strongly influenced by the time taken to construct a typical floor of the building, and here structural steel has an advantage over in-situ concrete. Even more time can be saved if the floor slabs are cast on permanent formwork that also acts as bottom reinforcement. These composite slabs, using corrugated steel sheeting as the formwork, have long been used in tall buildings in North America.[1] The relevant design methods are described in Section 3.10. In Europe, precast concrete permanent formwork and full-thickness precast floor slabs have also been found to be economical (Section 3.6.8), particularly in multi-storey car parks.[2]

At a time of rapid inflation and high interest rates, no comparison of the relative costs of different types of structure is meaningful unless account is taken of construction times. An example of this is provided by a 29-storey office block that was completed in London in 1974.[3] The structure consists of a slip-formed reinforced concrete core surrounded by a steel frame. This was designed to act compositely with the floor slabs, for which precast concrete planks and in-situ topping were used. On paper, this structure was more expensive than a reinforced concrete frame, but its construction time of 27 months was 8 months less than that for the concrete alternative. The consequent reduction in the total cost of the scheme justified the adoption of the composite structure.

The degree of fire protection that must be provided is another factor that influences the choice between concrete, composite, and steel structures, and here concrete has an advantage. Little or no fire protection is required for multi-storey car parks, a moderate amount for office blocks, more for public buildings, and most of all for multi-storey warehouses. Many methods have been developed for providing steelwork with fire protection, and information on them is readily available.[4, 5] Encasement in concrete is an economical method for steel columns, since the casing provides a substantial gain in strength. Most of the methods used for composite beams rely on lightweight materials, but in a few jobs the steel beams have been encased in concrete before erection. The encasement contributes little to the strength of the beam, but requires light-gauge reinforcement to control the width of cracks (Section 4.6) and to hold it in place during a fire.

The choice between steel, concrete and composite construction for a particular structure thus depends on many factors that are outside the scope of this book. But it is clear from recent practice that composite construction is particularly competitive in medium or long span structures where a concrete slab is needed for other reasons, where fire protection of steelwork is not required, and where there is a

Plate 1. Construction of hot mill floor for steel works, Newport, Monmouthshire (courtesy British Steel Corporation)

Plate 2. Rigid-jointed composite frame for north wing of the Baker Building, Engineering Laboratories, University of Cambridge

Plate 3. Use of precast planks and castella beams for floor at St Thomas's Hospital Medical School, London (courtesy Mr Gwylon Isaac)

Plate 4. Detail of precast slab and transverse reinforcement for Krupp-Montex flooring system (courtesy Fried. Krupp GmbH, Krupp Industrie- und Stahlbau, West Berlin)

premium on rapid construction. The use of composite floor structures in buildings of four different types is illustrated in Plates 1 to 4.

1.2 Composite columns and frames

When the stanchions in steel frames were first cased in concrete to protect them from fire, they were still designed for the applied load as if uncased. Then engineers realised that the encasement reduced the effective slenderness of a stanchion, and so increased its buckling load. Empirical methods for calculating the reduced slenderness are now given in design specifications for structural steelwork (Section 5.2).

This approach, although simple, is not rational, for the concrete encasement also carries its share of the load. Methods are now available for designing cased stanchions as composite columns. These take account of the true behaviour of these members, as found in tests to failure, and are described in Section 5.3.

When fire protection for the steel is not required, a composite column can be constructed without the use of formwork by filling a steel tube with concrete. Research on filled tubes led to their use in 1966 in a four-level motorway interchange at Almondsbury, near Bristol.[6] A design method is given in Section 5.4.

In framed structures, there may be composite beams, composite columns, or both. In developing design methods that take account of the interaction between beams and columns, it is necessary to consider many types of beam–column joint, and also to reconcile the differences between the methods now in use for concrete frames and for steel frames.

Two buildings with rigid-jointed composite frames were built in Great Britain in the early 1960s, at Cambridge[7] and London.[8] In a sense, these were full-scale experiments, for university staff engaged in relevant research were involved in their design. Also, at Imperial College, London, composite columns were used in the lower half of an otherwise reinforced concrete-framed structure of about 15 storeys, in order to maintain a constant size of column over the whole height of the building; but no account of this design has been published. The present state of development of design methods for such structures is described in Sections 5.5 and 5.6.

1.3 Design philosophy

An essential part of the design process is to take account of the random nature of loading, the variability of materials, and the defects that occur in construction, in such a way that the probability of unserviceability or failure of the structure during its design life is reduced to an acceptably low level. Extensive study of this subject since about 1950 has led to the incorporation of the older 'safety factor' and 'load factor' design methods into a more comprehensive 'limit state' design philosophy. Its first important application in Great Britain was in Code of Practice 110, *The Structural use of Concrete*, published in 1972.[9]

It is the policy of the British Standards Institution that future codes of practice and design specifications for structures shall be in limit-state form. Work began in 1968 on a 'Unified Bridge Code', to include steel, concrete, and composite superstructures and columns for bridges of all spans. The revision of BS 449[10] for steel structures in building began in 1969, and in 1970 it was decided to include composite steel–concrete structures within the new Standard, and to discontinue work on CP 117.[11, 12] The completion of the Bridge Code has been delayed by the need to take account in the 'steel' section of the lessons learned from several failures of box girders during the period 1968 to 1972, and from the research that followed. This subject is discussed further in Volume 2. It is hoped that the section of the new BS 449 on composite structures will be issued for public comment during 1976.

Limit-state design philosophy is used throughout this book, as this enables the principles and assumptions that underlie design methods to be set out more clearly than is possible in terms of the older philosophies. Many accounts of limit-state philosophy are available.[13, 14] The following brief summary of it is intended only to relate it to the older methods and to assist the reader to follow the design examples.

In traditional 'elastic' design, 'working' or 'permissible' stresses are obtained by dividing material strengths by a *safety factor* λ_e that has to take account of all types of uncertainty, including those associated with the loading. In the now well-established 'plastic' design method, the expected or 'working' loads are multiplied by a *load factor* λ_p that has to take account of variability of materials, as well as of loads. Limit-state design puts these factors where they rightly belong, by using two sets of *partial safety factors*, γ_f for loads and γ_m for materials.

Design calculations consist essentially of checking that a stress or stress resultant due to a given load does not exceed the strength of a given material or structural member. Figure 1.1 illustrates the fact that the numerical values at which these comparisons are made are different for the three design philosophies. The dots represent numbers fed into the calculation; the arrows show the effect of λ_e, λ_p, γ_m, and γ_f on these numbers, and the crosses mark the points where the checking is done.

The specified or calculated loads and the specified or guaranteed strengths of materials are known as *characteristic* values. After multiplication by γ_f or division by γ_m, as the case may be, they become *design* values. There is as yet no uniform practice for setting out these calculations. In CP 110 the relevant values of γ_m are often incorporated in numerical coefficients, so that *design* equations include symbols

such as f_{cu} and f_y, which represent *characteristic* strengths of the concrete and reinforcement, respectively. The same practice has been
followed for concrete in this book; but for steel, where γ_m values for
certain types of member are not yet established, symbols are used for
both characteristic and design strengths. The relationships are:

for reinforcement, f_{rd} (design) $= f_{ry}$ (yield) $\div \gamma_m$
for rolled sections, f_{sd} (design) $= f_y$ (yield) $\div \gamma_m$

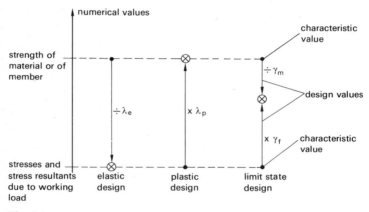

Fig. 1.1

Limit-state design is also more rational than the two older methods
in that it identifies more clearly the particular catastrophe or mode of
unserviceability that each part of the design procedure is intended to
avoid; and it provides a framework within which account can readily
be taken of the different degrees of risk, uncertainty, or variability
associated with various types of structure, loadings, or materials.

These situations when something 'goes wrong' are the *limit states*.
They fall into two groups, which can be called 'disasters' and
'nuisances'. Collapse, failure, overturning, buckling, or rupture of
part or all of a structure is a disaster, which may lead to loss of life.
These events are *ultimate limit states*. The relevant partial safety
factors all exceed 1·0, by amounts intended to ensure that the probability of such an occurrence is remote. Ultimate-strength and plastic
methods of analysis are appropriate, since inelastic behaviour usually
precedes failure; but elastic theory has to be used when no better
method is available.

Excessive deflection or vibration, unsightly cracks, and spalling of concrete, due perhaps to corrosion of reinforcement, are nuisances, in that they may require repair or may limit the usefulness of the structure, but they are not disasters. These are *serviceability limit states*. They normally occur while the structure is still elastic, so elastic analysis is appropriate. They must be avoided at working loads, so values of γ_f are usually 1·0. Their consequences are less severe than those of a failure, and to some extent (e.g., for deflections) they depend on average stiffnesses or strengths of materials, not on the occasional low value, so γ_m also can be taken as 1·0.

One situation that does not fit into this classification is fatigue failure under repeated loading. Initially a small fatigue crack is a form of unserviceability, but if allowed to spread it may eventually cause an ultimate limit state to be reached.

Limit-state design philosophy has been criticised on the ground that as the two limit states occur at different load levels, two sets of design calculations are needed, whereas with the older methods one was sufficient. This is only partly true, for it has been found possible when drafting codes of practice to identify many situations in which design for, say, the ultimate limit state will automatically ensure that certain types of unserviceability will not occur; and vice versa. In CP 110 the use of limiting span–depth ratios to control deflections and bar spacings to control cracking has made it unnecessary to analyse the structure at all for the serviceability loads. The development of similar design methods for composite structures is described in Sections 3.7, 4.5, and 4.6.

1.4 Properties of materials and shear connectors

Information on the properties of structural steel, concrete, and reinforcement is readily available, and will be given here only when it has particular relevance to composite structures. At low stresses linear elastic theory can be used for all three materials, although allowance must be made for the shrinkage and creep of concrete if accurate values of stress are required.

Design for the ultimate limit state is much influenced by the fundamental difference between the stress–strain curve for concrete in compression, and those for either type of steel, in tension or compression, which is illustrated in Fig. 1.2. Concrete reaches its maximum compressive stress at a strain of between 0·002 and 0·003, and at higher strains it crushes, losing almost all its compressive strength. It is very brittle in tension, having a strain capacity of only about 0·0001 (i.e., 0·1 mm per metre) before it cracks. The Figure also shows that the maximum stress reached by concrete in a beam or column is little more than 80% of its cube strength. Steel yields at a strain similar to that given for crushing of concrete, but on further straining the stress in steel continues to increase slowly, until the total strain is at least 40 times the yield strain. The subsequent necking and fracture is of no significance for composite members, for the useful strength of a cross-section is reached when the concrete crushes or the whole of the steel yields.

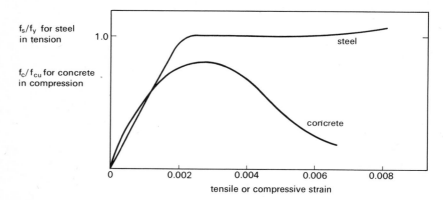

Fig. 1.2

Another difference is that the cube strength of concrete is more variable than the yield stress of steel. This is reflected in the values of the partial safety factors γ_m. In CP 110, these are based on the assumptions that the characteristic strengths of materials are the values

Table 1.1

	Concrete	Reinforcement
(1) Ultimate limit state, generally	1·5	1·15
(2) As (1), but for excessive loads and local damage	1·3	1·0
(3) Serviceability limit stage, generally	1·0	1·0
(4) As (3), but for stresses and crack widths at a cross-section of a member	1·3	1·0

below which not more than 5% of the test results fall, and that sufficient test data are available for statistical methods to be applicable.
The recommended values of γ_m are as shown in Table 1.1.

The reduced values in line (2) take account of the reduced probability
of the types of misuse or accident to which they apply. The values in
line (3) are appropriate for limit states where the average properties
of the member are relevant; but stresses or cracking at a particular
cross-section depend on the local properties of the concrete, so γ_m is
increased from 1·0 to 1·3 because local weakness is more probable
than weakness throughout a member.

These values of γ_m are used in the examples given here, with structural steel treated in the same way as reinforcement. This is in accordance with the current draft of the Bridge Code;[15] but a lower value
than 1·15, possibly 1·07, may be adopted for structural steel members
in the revised BS 449.

The only 'material' peculiar to composite construction is the shear
connector. Details of these and the measurement of their strength are
given in Chapter 2.

1.5   Loading

The main types of loading that arise in the design of structures for
buildings are now discussed. Loadings for bridge decks are considered
in Volume 2.

Dead load is that which is applied continuously during the life of
the structure. In composite members, the structural steel component
is usually built first, so it is necessary to distinguish between load
resisted by the steel component only, and load applied to the member
after the concrete has developed sufficient strength for composite
action to be effective. The division of the dead load between these
categories depends on the method of construction. Composite beams
are normally classified as *propped* or *unpropped*. In propped construction, the steel beam is supported at intervals along its length until the
concrete slab has reached a certain proportion, usually three-quarters,
of its design strength. The whole of the dead load is then assumed to be
resisted by the composite member. When no props are used, it is
assumed in elastic analysis that the steel member alone resists its own
weight and that of the formwork and the concrete slab. Other dead

loads, such as floor finishes and internal walls, are added later, and so are assumed to be carried by the composite member. In ultimate-strength methods of analysis (Section 3.4), it can be assumed that the effect of the method of construction on the strength of a member is negligible.

Dead loads can be determined quite accurately from known dimensions of the structure and densities of materials. These calculated weights are assumed to be the 'characteristic' loads.

Live (or imposed) loads for buildings are given in CP 3, Chapter 5.[16] These are essentially the 'working' loads used in earlier design methods, for it is not yet possible to express loads in statistical terms, and these are now taken to be the 'characteristic' loads. The principal loading in buildings is a uniformly distributed load on each floor, with the proviso that any small area 0·3 m square must be capable of carrying a much higher load.

Wind loads for buildings are also given in CP 3, Chapter 5. These usually consist of a uniform pressure or suction on each external surface. They rarely influence the design of composite beams, but can be important in framed structures not braced against side-sway (Section 5.6).

Thus methods of calculation that consider point and distributed loads are sufficient for all types of external loading. Shrinkage and creep of concrete and differential changes of temperature can all cause internal stress in composite structures, and so have to be considered as a form of loading. Their effect on stresses and deflections in beams is considered in Section 3.8.

The partial safety factors for dead and imposed loads given in CP 110 and likely to be adopted in the revised BS 449 are shown in Table 1.2.

Table 1.2

	Dead	Imposed
(1) Ultimate limit state, generally	1·4	1·6
(2) As (1), but for excessive loads	1·05	1·05
(3) Serviceability limit state	1·0	1·0

The 'excessive loads' in line (2) are those that might arise through misuse or accident. CP 110 also gives values for other load combinations that include wind loading.

1.6   Methods of analysis and design

Those who study composite structures usually have previous know-ledge of methods of analysis and design for either steel or concrete structures; but few are equally familiar with both materials. Con-tractors have the same problem; they are either 'steel erectors' or 'building contractors'. There are signs that this degree of specialisation, in both designers and constructors, is the chief limitation to the more extensive use of composite steel–concrete structures in buildings. It seems to be less of a problem in medium-span bridges, where com-posite structures are commonly used.

The purpose of this section is to provide a preview of the principal methods of analysis used in this book, and to show that most of them are straightforward applications of methods already in common use for steel or for concrete structures.

The steel designer will be familiar with the elementary elastic theory of bending, and the simple plastic theory in which the whole cross-section of a member is assumed to be at yield, either in tension or compression. Both theories are used for composite members, the differences being as follows:

(1) concrete in tension is neglected;
(2) in the elastic theory, concrete in compression is 'transformed' to steel by dividing its breadth by the modular ratio E_s/E_c;
(3) in the plastic theory, the equivalent 'yield stress' of concrete in compression can be assumed to be $0.4 f_{cu}$, where f_{cu} is the charac-teristic cube strength. Examples of this method are given in Sections 3.4, 4.3, and in Chapter 5.

The concrete designer will be familiar with the method of trans-formed sections, used extensively in Sections 3.7 and 4.5, and with the rectangular-stress-block theory outlined in (3) above. The basic difference from the elastic behaviour of reinforced concrete beams is that the steel section in a composite beam is more than 'tension reinforcement'. Account must be taken of the variation of stress across its depth; in other words, its flexural rigidity cannot be neglected.

The formulae for the elastic properties of composite members look more complex than those for either steel or reinforced concrete members. The chief reason is that the neutral axis for bending may lie in the web, the steel flange, or the concrete flange of the member. The theory is not in principle any more complex than that used for a steel I-beam. Values of the elastic section properties of composite beams are available from tables.[17,18]

Longitudinal shear.   Students usually find this subject troublesome, even though the formula

$$v = VA\bar{y}/Ib \qquad (1.1)$$

is familiar from their study of vertical shear stress in elastic beams, so a note on the use of this formula may be helpful. Its proof can be found in any undergraduate-level textbook on strength of materials.

We consider first the shear stresses in the elastic I-beam shown in Fig. 1.3, due to a vertical shear force V. For the cross-section 1–2 through the web, the 'excluded area' is the flange, of area A_f, and the

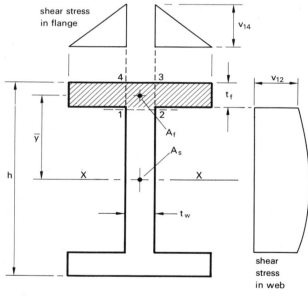

Fig. 1.3

distance $\bar{y}$ of its centroid from the neutral axis is $\frac{1}{2}(h - t_f)$. The longitudinal shear stress v_{12} on plane 1–2, of breadth t_w, is therefore

$$v_{12} = \tfrac{1}{2}VA_f(h - t_f)/It_w$$

where I is the second moment of area of the section about the axis
XX.

Consideration of the longitudinal equilibrium of the small element
1234 shows that if its area $t_w t_f$ is much less than A_f, then the mean
shear stress on planes 1–4 and 2–3 is given approximately by

$$v_{14} t_f = \tfrac{1}{2} v_{12} t_w$$

Repeated use of (1.1) for various cross-sections shows that the variation of longitudinal shear stress is parabolic in the web and linear in
the flanges, as shown in Fig. 1.3.

The second example is the elastic beam shown in section in Fig. 1.4.

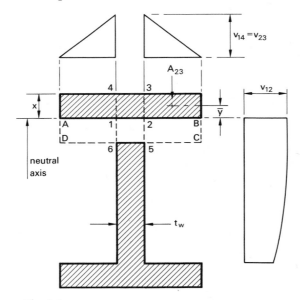

Fig. 1.4

This is similar to a composite beam with the neutral axis at depth x,
so that the cracked concrete in area $ABCD$ resists no longitudinal
bending stress. Here we assume the shaded area to be all of one
material. The 'gap' 1256 makes no difference to the applicability of
(1.1), because we assume that this area can transfer shear stress. Thus
the mean longitudinal shear stress on plane 2–3 through the flange is
given by

$$v_{23} = V A_{23} \bar{y} / I x$$

where A_{23} is the 'excluded area' and $\bar{y}$ is as shown in Fig. 1.4. Similarly, the variation of shear stress in the web is as shown. In the region 1256, the shear force $v_{12}t_w$ per unit length of beam is a more useful concept than the shear stress v_{12}, for this is the force resisted by the shear connection, according to elastic theory.

These results are used in the design of shear connectors for bridge girders; and the design of webs and of transverse reinforcement in flanges is based on distributions of shear stress like those shown in Fig. 1.4. But the elastic theory is not used in calculations for shear connectors for beams in buildings, as there is a simpler ultimate-strength method (Section 3.5).

Longitudinal slip.   Shear connectors are not rigid, so that a small longitudinal slip occurs between the steel and concrete components of a composite beam. This problem does not arise in other types of structure, and the relevant analysis is quite complex (Sections 2.6, 2.7, and Appendix A). An understanding of this subject is essential to anyone concerned with teaching or research in composite structures; but no detailed knowledge of it is necessary for design or construction, for all the methods of calculation used in practice have been developed in a way that enables slip to be neglected.

Deflections.   The effects of creep and shrinkage make the calculation of deflections in reinforced concrete beams more complex than for steel beams; but the limiting span–depth ratios given in CP 110 provide a simple means of checking that the deflection of a beam should not be excessive. The calculations for composite beams are similar (Sections 3.7 and 4.5.5) except that account must be taken of the method of construction of the beam.

Vertical shear.   The methods used for steel beams are applicable to all composite beams, but in continuous beams some economy can be gained by using the ultimate-strength method for composite beams that is given in Section 4.3.4.

Buckling of flanges and webs of beams.   This will be a new problem to most designers of reinforced concrete. In continuous beams it leads to restrictions on the slenderness of unstiffened flanges and webs (Section

4.4.2) that are more restrictive than those currently in use for steel beams. It is suggested in Section 5.5 that when steel members of slender cross-section are used in beams, semi-rigid beam–column joints may be the best method of achieving continuity.

Crack-width control.   The maximum spacings of reinforcing bars recommended in CP 110 are determined by the need to limit the width of visible cracks in reinforced concrete, for reasons of appearance and corrosion. Cracking is likely to be a problem in composite structures for buildings only in encased beams, or where the top surfaces of continuous beams are exposed to corrosion. The principles of crack-width control are the same as for reinforced concrete, and design formulae applicable to composite beams are given in Section 4.6. The rules for reinforcement-bar spacing, now being developed, are likely to be similar to those in CP 110.

Continuous beams.   In developing a simple design method for continuous beams in buildings (Chapter 4), use has been made of the simple plastic theory (as used for steel structures) and of redistribution of moments (as specified in CP 110 for concrete structures).

Columns and frames.   There is little similarity between the current design methods for steel frames (BS 449) and for concrete frames (CP 110). Composite members normally form part of a frame that is essentially steel, rather than concrete, so that the tentative design methods given in Chapter 5 are based on the established design methods for 'simple' and 'rigid-jointed' steel frames. For rigid frames, a method of elastic analysis of limited sub-frames is suggested, but it differs from that given in CP 110 in that allowance is made for plastic hinges at the ends of beams.

The economy of steel frames depends very much on the cost of the beam–column joints, and design methods have to be related to various types of joint in common use. It is this which makes it impracticable to recommend a single method for all types of composite frame. In buildings, semi-rigid joints (p. 180) could prove to be more economical than either simple or rigid joints, and research on this subject continues.

CHAPTER TWO

Shear Connection

2.1 Introduction

The established design methods for reinforced concrete and for structural steel provide answers to many of the problems that arise in the design of composite structures, but give no help with the basic problem of connecting the steel to the concrete. The force applied to this connection is mainly, but not entirely, longitudinal shear. As with bolted and welded joints, the connection is a region of severe and complex stress that defies accurate analysis, and so methods of connection have been developed empirically and verified by tests. They are described in Section 2.4.

The simplest type of composite member used in practice occurs in floor structures of the type shown in Fig. 3.1 (p. 42). The concrete floor slab is continuous over the steel I-sections, and is supported by them. It is designed to span in the y-direction in the same way as when supported by walls or the rib of a reinforced concrete T-beam. When shear connection is provided between the steel member and the concrete slab, the two together span in the x-direction as a composite beam. The steel member has not been described as a 'beam', because its main function at midspan is to resist tension, as does the reinforcement in a T-beam. The compression is assumed to be resisted by an 'effective' breadth of slab, b, as explained in Section 3.4.

Effective cross-sections of typical composite beams are therefore as shown in Fig. 2.1. The ultimate-strength design method used for such beams in buildings is described in Section 3.5. When shear connectors have to resist repeated loading, as in bridge girders, a method based on elastic theory is used. This will be explained in Volume 2.

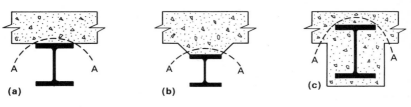

Fig. 2.1

The purpose of the present chapter is to describe the effects of shear connection on the behaviour of very simple beams, and hence to show how current methods of shear connection and design procedures for beams were developed.

2.2   Simply supported beam of rectangular cross-section

Flitched beams, whose strength depended on shear connection between parallel timbers, were used in mediaeval times, and survive today in the form of glued-laminated construction. Such a beam, made from two members of equal size (Fig. 2.2), will now be studied. It carries a load w per unit length over a span L, and its components are made of an elastic material with Young's modulus E. The weight of the beam is neglected.

2.2.1   No shear connection

We assume first that there is no shear connection or friction on the interface AB. The upper beam cannot deflect more than the lower one, so each carries load $w/2$ per unit length as if it were an isolated beam of second moment of area $bh^3/12$, and the vertical compressive stress across the interface is $w/2b$. The midspan bending moment in each beam is $wL^2/16$. By elementary beam theory, the stress distribution at midspan is as in Fig. 2.2(c), and the maximum bending stress in each component, f, is given by

$$f = \frac{My_{max}}{I} = \frac{wL^2}{16}\frac{12}{bh^3}\frac{h}{2} = \frac{3wL^2}{8bh^2} \tag{2.1}$$

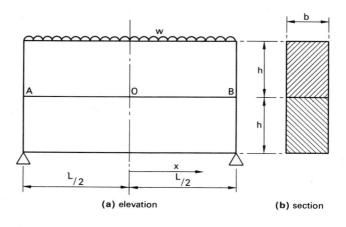

(a) elevation   (b) section

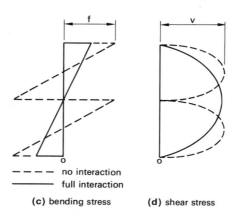

- - - - no interaction
——— full interaction

(c) bending stress   (d) shear stress

Fig. 2.2

The maximum shear stress, v, occurs near a support. The parabolic distribution given by simple elastic theory is shown in Fig. 2.2(d); and at the centre-line of each member,

$$v = \frac{3}{2}\frac{wL}{4}\frac{1}{bh} = \frac{3wL}{8bh} \tag{2.2}$$

The maximum deflection, δ, is given by the usual formula

$$\delta = \frac{5(w/2)L^4}{384EI} = \frac{5}{384}\frac{w}{2}\frac{12L^4}{Ebh^3} = \frac{5wL^4}{64Ebh^3} \tag{2.3}$$

The bending moment in each beam at a section distant x from midspan is $M_x = w(L^2 - 4x^2)/16$, so that the longitudinal strain ε_x at the bottom fibre of the upper beam is

$$\varepsilon_x = \frac{My_{\max}}{EI} = \frac{3w}{8Ebh^2}(L^2 - 4x^2) \tag{2.4}$$

There is an equal and opposite strain in the top fibre of the lower beam, so that the difference between the strains in these adjacent fibres, known as the *slip strain*, is $2\varepsilon_x$.

It is easy to show by experiment with two or more flexible wooden laths or rulers that under load, the end faces of the two-component beam have the shape shown in Fig. 2.3(a). The slip at the interface, s, is zero at $x=0$ (from symmetry) and a maximum at $x = \pm L/2$. The cross-section $x = 0$ is the only one where plane sections remain plane. The slip strain, defined above, is not the same as slip. In the same way that strain is rate of change of displacement, slip strain is the rate of change of slip along the beam. Thus from (2.4),

$$\frac{ds}{dx} = 2\varepsilon_x = \frac{3w}{4Ebh^2}(L^2 - 4x^2) \tag{2.5}$$

Integration gives

$$s = \frac{w}{4Ebh^2}(3L^2x - 4x^3) \tag{2.6}$$

The constant of integration is zero, since $s = 0$ when $x = 0$, so that (2.6) gives the distribution of slip along the beam.

Results (2.5) and (2.6) for the beam studied in Section 2.7 are plotted in Fig. 2.3. This shows that at midspan, slip strain is a maximum and slip is zero, and at the ends of the beam, slip is a maximum and slip strain is zero. From (2.6), the maximum slip (when $x = L/2$) is $wL^3/4Ebh^2$. Some idea of the magnitude of this slip is given by relating it to the maximum deflection of the two beams. From (2.3), the ratio of slip to deflection is $3.2h/L$. The ratio $L/2h$ for a beam is typically about 20, so that the end slip is less than a tenth of the deflection. We conclude that *shear connection must be very stiff if it is to be effective.*

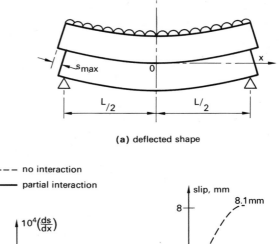

(a) deflected shape

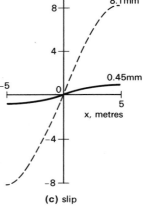

(b) slip strain

(c) slip

Fig. 2.3

2.2.2 Full interaction

It is now assumed that the two halves of the beam shown in Fig. 2.2 are joined together by an infinitely stiff shear connection. The two members then behave as one. Slip and slip strain are everywhere zero, and it can be assumed that plane sections remain plane. This situation is known as *full interaction*. With one exception (Section 3.9), all design of composite beams and columns in practice is based on the assumption that full interaction is achieved.

For the composite beam of breadth b and depth $2h$, $I = 2bh^3/3$, and elementary theory gives the midspan bending moment as $wL^2/8$. The extreme fibre bending stress is

$$f = \frac{My_{max}}{I} = \frac{wL^2}{8} \frac{3}{2bh^3} h = \frac{3wL^2}{16bh^2} \tag{2.7}$$

The vertical shear at section x is

$$V_x = wx \tag{2.8}$$

so the shear stress at the neutral axis is

$$v_x = \tfrac{3}{2}wx \frac{1}{2bh} = \frac{3wx}{4bh} \tag{2.9}$$

and the maximum shear stress is

$$v = 3wL/8bh \tag{2.10}$$

These stresses are compared in Figs. 2.2(c) and (d) with those for the non-composite beam. Due to the provision of the shear connection, the maximum shear stress is unchanged, but the maximum bending stress is halved.

The midspan deflection is

$$\delta = \frac{5wL^4}{384EI} = \frac{5wL^4}{256Ebh^3} \tag{2.11}$$

which is one-quarter of the previous deflection (Eq.(2.3)). Thus the provision of shear connection increases both the strength and the stiffness of a beam of given size, and in practice leads to a reduction in the size of beam required for a given loading, and usually to a reduction in its cost.

In this example—but not always—the interface AOB coincides with the neutral axis of the composite member, so that the maximum longitudinal shear stress at the interface is equal to the maximum vertical shear stress, which occurs at $x = \pm L/2$ and is $3wL/8bh$, from (2.10). This example suggests what is found to be true in practice: that *the shear connection needs to be as strong in shear as the weaker of the two materials joined.*

The shear connection must be designed for the longitudinal shear per unit length, q, which is known as the *shear flow*. In this example it is given by

$$q_x = v_x b = 3wx/4h \qquad (2.12)$$

When design is on an elastic basis, the shear connectors are spaced in accordance with the shear flow. Thus if the design shear force per connector is P_d, the pitch or spacing, p, at which they should be spaced is given by $pq_x \not> P_d$, which is, from (2.12),

$$p \not> 4P_d h/3wx \qquad (2.13)$$

This is known as 'triangular' spacing, from the shape of the graph of q plotted against x (Fig. 2.4).

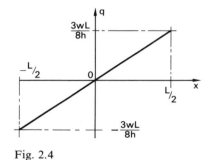

Fig. 2.4

2.3 Uplift

In the preceding example, the stress normal to the interface AOB (Fig. 2.2) was everywhere compressive, and equal to $w/2b$ except at the ends of the beam. The stress would have been tensile if the load w had been applied to the lower member. Such loading is unlikely, except when travelling cranes are suspended from the steelwork of a composite floor above; but there are other situations in which stresses tending to cause uplift can occur at the interface. These arise from complex effects such as the torsional stiffness of reinforced concrete slabs forming flanges of composite beams, the triaxial stresses in the vicinity of shear connectors, and, in box-girder bridges, the torsional stiffness of the steel box.

Tension across the interface can also occur in beams of non-uniform section or with partially completed flanges. Two members without shear connection, as shown in Fig. 2.5, provide a simple example. *AB* is supported on *CD* and carries distributed loading. It can easily be shown by elastic theory that if the flexural rigidity of *AB* exceeds about one-tenth of that of *CD*, then the whole of the load on *AB* is transferred to *CD* at points *A* and *B*, with separation of the beams between these points. If *AB* were connected to *CD*, there would be uplift forces at midspan.

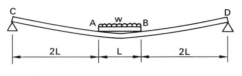

Fig. 2.5

Almost all connectors used in practice are therefore so shaped that they provide resistance to uplift as well as to slip. Uplift forces are so much less than shear forces that it is not normally necessary to calculate or estimate them for design purposes, provided that connectors with some uplift resistance are used.

2.4 Methods of shear connection

2.4.1 Bond

Until the use of deformed bars became common, most of the reinforcement for concrete consisted of smooth mild-steel bars. The transfer of shear from steel to concrete was assumed to occur by bond or adhesion at the concrete–steel interface. Where the steel component of a composite member is surrounded by reinforced concrete, as in an encased beam, Fig. 2.1(c), or an encased stanchion, Fig. 5.2, the analogy with reinforced concrete suggests that no shear connectors need be provided. Tests have shown that this is usually true for cased stanchions and filled tubes, where bond stresses are low, and also for cased beams in the elastic range. But in design it is necessary to restrict bond stress to a low value, to provide a margin for the incalculable effects of shrinkage of concrete, poor adhesion to the underside of steel surfaces, and stresses due to variations of temperature.

Research on the ultimate strength of cased beams[19] has shown that at high loads, calculated bond stresses have little meaning, due to the development of cracking and local bond failures. If longitudinal shear failure occurs, it is invariably on a plane such as AA in Fig. 2.1(c), and not around the perimeter of the steel section. For these reasons, British codes of practice do not allow ultimate-strength design methods to be used for composite beams without shear connectors.

Most composite beams have cross-sections of types (a) or (b) in Fig. 2.1. Tests on such beams show that at low loads, most of the longitudinal shear is transferred by bond at the interface, that bond breaks down at higher loads, and that once broken it cannot be restored. So in design calculations, bond strength is taken as zero, and in research, bond is deliberately destroyed by greasing the steel flange before the concrete is cast. For uncased beams, the most practicable form of shear connection is some form of dowel welded to the top flange of the steel member and subsequently surrounded by in-situ concrete when the floor or deck slab is cast.

2.4.2   Shear connectors

The most widely used type of connector is the headed stud (Fig. 2.6). These range in diameter from 13 to 25 mm, and in length (h) from 65 to 100 mm, though longer studs are sometimes used in haunched beams.

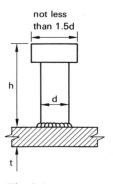

Fig. 2.6

British specifications for composite structures for buildings and bridges require them to be made from steel with a minimum elongation of 18% and characteristic yield stress not less than 400 N/mm^2, and to be attached to the steel member by an automatic stud-welding tool. Their advantages are that the welding process is rapid and simple, they provide little obstruction to reinforcement in the concrete slab, and they are equally strong and stiff in shear in all directions normal to the axis of the stud.

There are two factors that influence the diameter of studs. One is the welding process, which becomes increasingly expensive and difficult at diameters exceeding 19 mm, and the other is the thickness t (Fig. 2.6) of the plate or flange to which the stud is welded. A study made in the USA[20] found that the full static strength of the stud can be developed if d/t is less than about 2·7, and a limit of 2·0 is used in Britain. Tests using repeated loading[21] led to the rule that where the flange plate is subjected to fluctuating tensile stress, d/t may not exceed 1·5.

Thus the maximum shear force that can be resisted by a stud is

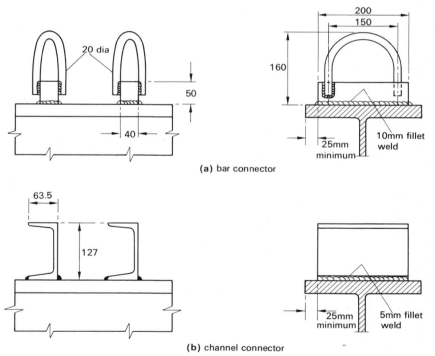

(a) bar connector

(b) channel connector

Fig. 2.7

relatively low (about 150 kN). Other types of connector with higher strength have been developed, primarily for use in bridges. These are bars with hoops (Fig. 2.7(a)), tees with hoops, horseshoes and channels (Fig. 2.7(b)). The largest sizes specified are shown. Bars with hoops are the strongest of these, with ultimate shear strengths of up to 1000 kN.

These types of shear connector are not ideal for use with decks or floors composed of precast concrete units. Such units have been connected to steel beams by means of high-strength friction-grip bolts, but the method is not widely used. Epoxy adhesives have also been tried, but it is not clear how resistance to uplift can reliably be provided when the slab is attached to the steel member only at its lower surface.

2.4.3   Corrugated metal decking

In North America, suspended floors for buildings are commonly constructed by casting a concrete slab on permanent formwork consisting of corrugated steel decking. This practice is becoming more common in Great Britain, where the material is also known as *profiled steel sheet*. Typical cross-sections of floors of this type are shown in Figs. 2.8 and 3.21. It is impracticable to weld shear connectors to

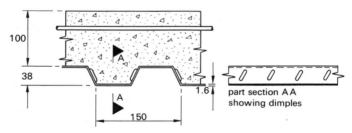

Fig. 2.8

material that may be only 1 mm thick, so shear connection is provided by bond at the steel–concrete interface, and in some sections also by pressed or rolled dimples that project from the sides of the corrugations into the concrete. Uplift is prevented either by the shape of the profile (Fig. 3.21) or by inclining the dimples to the vertical in opposite directions on the two sides of a corrugation (Fig. 2.8).

Several proprietary floor systems of these types are available. Their design and their use with composite beams is considered in Section 3.10.

2.5   Tests on shear connectors

The property of a shear connector of most relevance to design is the relationship between the shear force transmitted, P, and the slip at the interface, s. This load-slip curve should ideally be found from tests on composite beams, but in practice a simpler specimen is necessary. Most of the data on connectors have been obtained from various types of 'push-out' tests. The flanges of a short length of steel beam are connected to two small concrete slabs, as shown in Fig. 2.9.

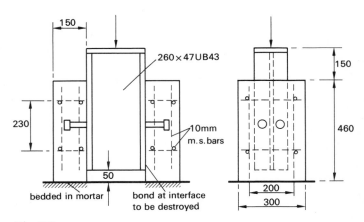

Fig. 2.9

The slabs are bedded in mortar on the lower platen of a standard compression-testing machine, and load is applied to the upper end of the steel beam. Slip between the beam and the two slabs is measured by dial gauges, and the average slip is plotted against the load per connector. A typical load–slip curve is shown in Fig. 2.10.

In practice, designers normally specify shear connectors for which strengths have already been established, for it is an expensive matter to carry out the tests needed to establish a full set of design strengths for a new type of connector. Study of the push-out test as standardised in British codes of practice[11] shows the reason for this. If reliable results are to be obtained, the test must be specified in detail, for the load–slip relationship is influenced by many variables, including

(1) number of connectors in the test specimen,
(2) mean longitudinal stress in the concrete slab surrounding the connectors,
(3) size, arrangement, and strength of slab reinforcement in the vicinity of the connectors,
(4) thickness of concrete surrounding the connectors,
(5) freedom of the base of each slab to move laterally, and so to impose uplift forces on the connectors,
(6) bond at the steel–concrete interface,
(7) strength of the concrete slab, and
(8) degree of compaction of the concrete surrounding the base of each connector.

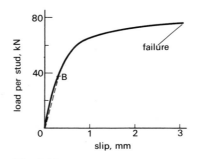

Fig. 2.10

The details shown in Fig. 2.9 include requirements relevant to items 1 to 6 above. The specified amount of reinforcement is less than that which would normally be used in a beam, so that in this respect, strengths found from the test are likely to be on the safe (low) side. The strength of the concrete can influence the mode of failure, as well as the failure load. In this test, large-diameter studs reach their maximum load when the slab surrounding them fails, but small-diameter studs shear off. The change of mode occurs at a diameter of from 16 to 19 mm, depending on the concrete strength.

Typical values for characteristic static strength for headed studs are given in Table 2.1. These show that the strength is roughly proportional to the square of the diameter and to the square root of the concrete strength.

Table 2.1

Headed studs		Static strength in kN for characteristic concrete cube strengths f_{cu} in N/mm^2			
Diameter mm	Height mm	20	30	40	50
25	100	139	154	168	183
19	100	90	100	109	119
13	65	42	47	52	57

Research has shown that the bearing stress on the shank of a stud connector varies along the length as sketched in Fig. 2.11. To obtain a rough estimate of the maximum bearing stress, let us assume that the load given in Table 2.1 is carried by a length of connector equal to twice its diameter. For a 19-mm stud in concrete with a cube strength of 40 N/mm^2, the mean stress on this loaded area is found to be four times the cube strength. A similar calculation for a bar-and-hoop connector, averaging the specified load over the area of the bar, gives a mean bearing stress of 2·5 times the cube strength, which confirms that the previous result is of the right order.

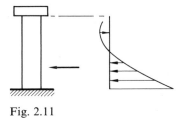

Fig. 2.11

These very high stresses are possible only because the concrete bearing on the connector is restrained laterally by the surrounding concrete, its reinforcement, and the steel flange. The results of push-out tests are likely to be influenced by the degree of compaction of the concrete, and even by the arrangement of particles of aggregate, in this small but critical region, and this is thought to be the main reason for the scatter of results obtained. The usual way of allowing for this scatter is to specify that the characteristic strength P_k be taken as the lowest of the failure loads P_u (after correction for any difference between the strength of the concrete and the specified value) found from tests on three supposedly identical specimens.

The load–slip curve for a connector in a beam cannot be assumed to be that found in a push-out test, for the distribution of longitudinal stress in the concrete flange of a beam is different from that in the slab in a push-out test. In simply supported beams, where the flange is in compression, the stiffness of the shear connection (i.e., the ratio of load to slip) in the elastic range may be up to twice as great as that found in a push-out test, but the ultimate strength is about the same. It has been found[22] that when the slab is in tension the connection is less stiff, and the ultimate strength is slightly reduced. It is therefore being proposed in the draft code of practice for composite beams in buildings[23] that the ultimate strength of connectors in negative-moment regions of continuous beams should be taken as 20% less than the value used for positive-moment regions.

There are three other situations in which a connector strength found in a push-out test may be too high for use in design. The most important is repeated loading, such as that due to the passage of traffic over a bridge. Codes of practice for bridge design give appropriate reduced loads for connectors, based on the results of fatigue tests on push-out specimens and on composite beams.

Loss of strength may also be expected when the lateral restraint to the concrete in contact with the connector is less than that provided in a push-out test, as in a haunched beam with connectors too close to a free surface (Fig. 2.12). For this reason, the use of the tabulated

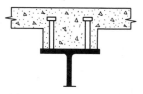

Fig. 2.12

connector strengths in haunched beams is allowed only when the cross-section of the haunch satisfies empirical rules deduced from the limited test data at present available. The two rules proposed for beams in buildings are shown in Fig. 2.13. The concrete cover to the side of the connector may not be less than 40 mm (line AB), and the free concrete surface may not lie within the line CD, which runs from the base of the connector at an angle of 45° with the steel flange. Thus EFG shows the smallest haunch satisfying these rules. A designer can use a more slender haunch if appropriate design loads for the shear connectors are determined by push-out tests on specimens with haunches of the shape proposed.

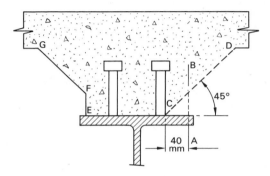

Fig. 2.13

There are also rules for the detailing of haunch reinforcement, which are discussed in Section 3.6.2. These are likely to determine the profile of the slab at the free edge of an L-beam.

Tests show that the ability of lightweight-aggregate concrete to resist the very high local stresses at shear connectors is slightly less than that of normal-density concrete of the same cube strength. It is usual to allow the use of this material in composite beams, provided that its density exceeds 1400 kg/m³ (60% of that of normal-density concrete), and that the connector strengths are taken as 15% less than the tabulated values.

2.6   Partial interaction

In studying the simple composite beam with full interaction (Section 2.2.2), it was assumed that slip was everywhere zero. But the results of push-out tests show (e.g., Fig. 2.10) that even at the smallest loads, slip is not zero. It is therefore necessary to know how the behaviour of a beam is modified by the presence of slip. This is best illustrated by an analysis based on elastic theory. It leads to a differential equation that has to be solved afresh for each type of loading, and is therefore too complex for use in design offices. Even so, partial-interaction theory is useful, for it provides a starting point for the development of simpler methods for predicting the behaviour of beams at working load, and finds application in the calculation of interface shear forces due to shrinkage and differential thermal expansion.

The problem to be studied and the relevant variables are defined below. The details of the theory, and of its application to a composite beam, are given in Appendix A. The results and comments on them are given below and in Section 2.7.

Elastic analysis is relevant to situations in which the loads on connectors do not exceed about half their ultimate strength. The relevant part OB of the load–slip curve (Fig. 2.10) can be replaced with little error by the straight line OB. The ratio of load to slip given by this line is known as the *connector modulus, k*.

For simplicity, the scope of the analysis is restricted to a simply supported composite beam of span L (Fig. 2.14), carrying a distributed

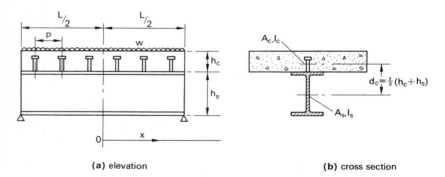

(a) elevation   (b) cross section

Fig. 2.14

load w per unit length. The cross-section consists of a concrete slab of thickness h_c, cross-sectional area A_c, and second moment of area I_c, and a symmetrical steel section with corresponding properties h_s, A_s, and I_s. The distance between the centroids of the concrete and steel cross-sections, d_c, is given by

$$d_c = (h_c + h_s)/2 \qquad (2.14)$$

Shear connectors of modulus k are provided at uniform spacing p along the length of the beam.

The elastic modulus of the steel is E_s, and that of the concrete for short-term loading is E_c. Allowance is made for creep of concrete by using an effective modulus E_c' in the analysis, where

$$E_c' = k_c E_c,$$

and k_c is a reduction coefficient, calculated from the ratio of creep strain to elastic strain. The modular ratio m is defined by $m = E_s/E_c$, so that

$$E_c' = k_c E_s/m \qquad (2.15)$$

The concrete is assumed to be as stiff in tension as in compression, for it is found that tensile stresses in concrete are low enough for little error to result in this analysis, except when the degree of shear connection is very low.

The results of the analysis are expressed in terms of two functions of the cross-section of the member and the stiffness of its shear connection, α and β. These are defined by the following equations, in which the notation established in CP117: Part 2[12] has been used.

$$\frac{1}{A_0} = \frac{m}{k_c A_c} + \frac{1}{A_s} \qquad (2.16)$$

$$\frac{1}{A'} = d_c^2 + \frac{I_0}{A_0} \qquad (2.17)$$

$$I_0 = \frac{k_c I_c}{m} + I_s \qquad (2.18)$$

$$\alpha^2 = k/(pE_s I_0 A') \qquad (2.19)$$

$$\beta = A' p d_c/k \qquad (2.20)$$

In a composite beam, the steel section is thinner than the concrete section, and steel has a much higher coefficient of thermal conductivity. Thus the steel responds more rapidly than the concrete to changes of temperature. If the two components were free, their length would change at different rates; but the shear connection prevents this, and the resulting stresses in both materials can be large enough to influence design. The shrinkage of the concrete slab has a similar effect. A simple way of allowing for such differential strains in this analysis is to assume that after connection to the steel, the concrete slab shortens uniformly, by an amount ε_c per unit length, relative to the steel.

It is shown in Appendix A that the governing equation relating slip s to distance along the beam from midspan, x, is

$$\frac{d^2 s}{dx^2} - \alpha^2 s = -\alpha^2 \beta w x \qquad (2.21)$$

and that the boundary conditions for the present problem are:

$$\left. \begin{array}{ll} s = 0 & \text{when } x = 0 \\ ds/dx = -\varepsilon_c & \text{when } x = \pm L/2 \end{array} \right\} \qquad (2.22)$$

The solution of (2.21) is then

$$s = \beta w x - \left(\frac{\beta w + \varepsilon_c}{\alpha} \right) \operatorname{sech} \left(\frac{\alpha L}{2} \right) \sinh \alpha x \qquad (2.23)$$

Expressions for the slip strain and the stresses throughout the beam can be obtained from this result. The stresses at a cross-section are found to depend on the loading, boundary conditions and shear connection for the whole beam. They cannot be calculated from the bending moment and shear force at the section considered. This is the main reason why design methods simple enough for use in practice have to be based on full-interaction theory.

2.7   Effect of slip on stresses and deflections

It is now assumed that the reader is familiar with the method of *transformed sections*, which was developed for the analysis of elastic members composed of two or more materials with different moduli of elasticity. Accounts of it will be found in elementary texts on reinforced concrete or strength of materials.

Full-interaction and no-interaction elastic analyses are given in Section 2.2 for a composite beam made from two elements of equal size and stiffness. Its cross-section (Fig. 2.2(b)) can be considered as the transformed section for the steel and concrete beam shown in Fig. 2.15. Partial-interaction analysis of this beam (Appendix A) illustrates well the effect of connector flexibility on interface slip and hence on stresses and deflections, even though the cross-section is not one that would be used in practice.

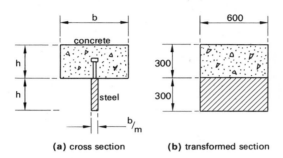

(a) cross section   (b) transformed section

Fig. 2.15

The numerical values, chosen to be typical of a composite beam, are given in Section A.2 (p. 194). Substitution in (2.23) gives the relation between s and x for a beam of depth 0·6 m and span 10 m as

$$10^4 s = 1·05x - 0·0017 \sinh (1·36x) \qquad (2.24)$$

The maximum slip occurs at the ends of the span, where $x = \pm 5$ m. From (2.24), it is $\pm 0·45$ mm.

The results obtained in Sections 2.2.1 and 2.2.2 are also applicable to this beam. From (2.6), the maximum slip if there were no shear connection would be $\pm 8·1$ mm. Thus the shear connectors reduce end slip substantially, but do not eliminate it. The variations of slip strain and slip along the span for no interaction and partial interaction are shown in Fig. 2.3.

The connector modulus k was taken as 150 kN/mm (Appendix A). The maximum load per connector is k times the maximum slip, so the partial-interaction theory gives this load as 67 kN, which is sufficiently far below the ultimate strength of 100 kN per connector for the assumption of a linear load–slip relationship to be reasonable. Longitudinal strains at midspan given by full-interaction and partial-interaction theory are shown in Fig. 2.16. The increase in extreme-fibre strain due to slip, 28×10^{-6}, is much less than the slip strain at the interface, 104×10^{-6}. The maximum compressive stress in the concrete is increased by slip from 12·2 to 12·8 N/mm², a change of 5%. This higher stress is 43% of the cube strength, so the assumption of elastic behaviour is reasonable.

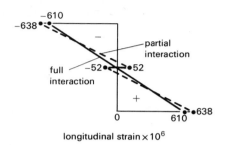

longitudinal strain × 10⁶

Fig. 2.16

The ratio of the partial-interaction curvature to the full-interaction curvature is 690/610, or 1·13. Integration of curvatures along the beam shows that the increase in deflection, due to slip, is also about 13%. The effects of slip on deflection are found in practice to be less than is implied by this example, because here a rather low value of connector modulus has been used, and the effect of bond has been neglected.

The longitudinal compressive force in the concrete at midspan is proportional to the mean compressive strain. From Fig. 2.16, this is 305×10^{-6} for full interaction and 293×10^{-6} for partial inter-action, a reduction of 4%.

The influence of slip on the flexural behaviour of the member may be summarised as follows. The bending moment at midspan, $wL^2/8$, can be considered to be the sum of a 'concrete' moment M_c, a 'steel' moment M_s, and a 'composite' moment Fd_c (Fig. A.1, p. 191):

$$M_c + M_s + Fd_c = wL^2/8$$

In the full-interaction analysis, Fd_c contributes 75% of the total moment, and M_c and M_s 12·5% each. The partial-interaction analysis shows that slip reduces the contribution from Fd_c to 72% of the total, so that the contributions from M_c and M_s rise to 14%, corresponding to an increase in curvature of $(14 - 12\cdot5)/12\cdot5$, or about 13%.

The interface shear force per unit length, q_x, is given by (2.12) for full interaction and by (A.1) and (2.24) for partial interaction. The expressions for q_x over a half span are plotted in Fig. 2.17, and show that in the elastic range, the distribution of loading on the connectors is similar to that given by full-interaction theory. The reasons for using uniform rather than 'triangular' spacing of connectors are discussed in Section 3.5.

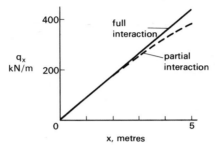

Fig. 2.17

CHAPTER THREE

Simply Supported Composite Beams and Slabs

3.1 Design methods

The existing British codes of practice for composite beams, CP 117: Part 1[11] and CP 117:Part 2,[12] are being revised, and will be incorporated in two new documents, a rewritten and expanded BS 449 (*The Use of Structural Steel in Buildings*)[23] and a new code or specification, *Steel, Concrete, and Composite Bridges*,[15] both of which are being drafted in accordance with limit-state design philosophy. The methods of analysis and design for composite structures likely to be recommended in the new BS 449 will be described in this and the following chapters, and illustrated by a worked example. To avoid repetition, the results obtained at each stage in this design example are carried forward and used in subsequent work. Methods that find application only in bridge decks will be covered in Volume 2.

Due to the changes in design philosophy, it will not be possible to use the new codes in conjunction with the existing BS 449, BS 153, or CP 117; but they are intended to be compatible with the only limit-state code so far available, CP 110, *The Structural Use of Concrete*,[9] which is used where appropriate in the design example.

The descriptions of structural behaviour that follow are related to established theories and published research, rather than to the two draft codes, for these are subject to alteration; but the details of the design calculations are generally in accordance with the current drafts (April 1974). It must not be assumed that the numerical values used (for example, of partial safety factors) will necessarily be adopted in the final versions.

3.2   The design example

In a framed structure for a wing of a building, the columns are arranged at 4 m centres in two rows 9 m apart. A design is required for a typical floor, which consists of a reinforced concrete floor slab continuous over and composite with steel beams that span between the columns as shown in Fig. 3.1. The characteristic material strengths and loads, and the partial safety factors at the ultimate limit state for materials (γ_m) and for loads (γ_f) are assumed to be as follows:

Structural steel; yield strength $f_y = 350 \text{ N/mm}^2$, $\gamma_m = 1\cdot15$
Reinforcement; yield strength $f_{ry} = 410 \text{ N/mm}^2$, $\gamma_m = 1\cdot15$
Concrete; cube strength $f_{cu} = 30 \text{ N/mm}^2$, $\gamma_m = 1\cdot5$
Shear connectors; strengths as in Table 2.1 $\gamma_m = 1\cdot15$
Dead load; as designed, assuming density of
    reinforced concrete to be 2400 kg/m^3; $\gamma_f = 1\cdot4$
Imposed load, finishes and partitions, etc., per
    unit area of typical floor slab; 7·5 kN/m^2, $\gamma_f = 1\cdot6$

At the serviceability limit state, all partial safety factors are taken as 1·0.

The design yield strengths of the structural steel (f_{sd}) and reinforcement (f_{rd}) at the ultimate limit state are given by

$$f_{sd} = 0\cdot87 f_y = 304 \text{ N/mm}^2 \qquad f_{rd} = 0\cdot87 f_{ry} = 356 \text{ N/mm}^2$$

Finishes and partitions have here been included with imposed load for simplicity. When details of the finishes and location of permanent non-structural partition walls are known when the structure is designed, these should strictly be treated as dead load. This slightly reduces the design load at the ultimate limit state, since γ_f is lower, but it increases the calculated deflection due to creep of concrete.

The modulus of elasticity of steel, E_s, is taken as 200 kN/mm^2. The static modulus of the concrete is given in Appendix D of CP 110 as ranging from 23 to 33 kN/mm^2. It is here assumed that $E_c = 26\cdot7$

kN/mm², giving a modular ratio for short-term loading of 7·5. When it is necessary to allow for creep, the factor k_c (p. 35) is taken as 0·5 (i.e., creep strain is assumed equal to elastic strain, as in CP 117: Part 2).

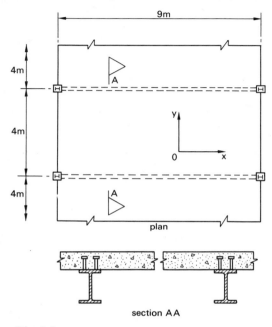

Fig. 3.1

Except where noted, the methods to be described are applicable to beams in bridges as well as to this simplified example of a beam in a building, but in bridge beams other problems not studied here, such as the effects of repeated or non-uniform loading, must also be considered in practice.

3.3   The floor slab

The usual methods for the design of one-way and two-way slabs are applicable also when the slab serves as the flange of one or more composite beams, but in this situation the designer should aim to provide as much flexural strength as possible in the negative (hogging) moment regions where the slab is supported on a steel member. The bending of the slab then provides lateral restraint at the level where the shear connectors impose the maximum bursting forces on the concrete (Fig. 3.2). It will be shown (p. 57) that more economical design of the reinforcement surrounding the connectors is then possible.

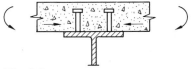

Fig. 3.2

When precast concrete floor or deck units are used, careful detailing and construction are needed to ensure adequate containment for the connectors. For example, if precast planks used as permanent formwork are detailed as in Fig. 3.3(a) but built as in Fig. 3.3(b), the in-situ concrete in the narrow gap between their ends is likely to be of poor quality, and the performance of the connectors will suffer.

The transverse joints between the sides of precast units (*AB* in Fig. 3.4) and the ends of the units themselves must both be designed to

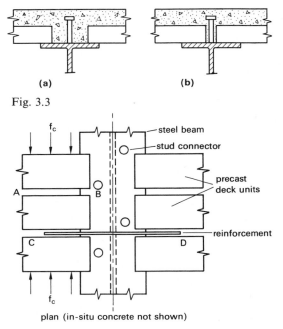

(a)   **(b)**

Fig. 3.3

Fig. 3.4

resist the compressive stresses f_c that occur when they form part of the flange of a composite beam. The requirements for bottom transverse reinforcement (p. 54) can be met by placing bars in the in-situ joints (*CD* in Fig. 3.4) if the units are narrow. The design of wide units is considered in Section 3.6.8.

3.3.1   Example. The floor slab

A one-way slab will suit the beam layout shown in Fig. 3.1. It will be designed to CP 110.[9] The span is 4 m and the slab thickness is assumed to be 150 mm, which gives a span–depth ratio that satisfies clause 3.3.8 of CP 110 in respect of deflection of the slab. The design loads are as shown in Table 3.1. The names used here for the limit states are

Table 3.1

	Serviceability limit state	γ_f	Ultimate limit state
Dead load; weight of slab			
$2.4 \times 9.81 \times 0.15$	3.6	× 1.4	5.0
Imposed load	7.5	× 1.6	12.0
Total design load for slab	$\overline{11.1} \text{ kN/m}^2$		$\overline{17.0} \text{ kN/m}^2$

in accordance with CP 110. In the draft codes for composite structures the terms *unserviceability limit state* and *collapse limit state* are also used. They may be assumed to have the same meanings as those given above.

At the ultimate limit state (i.e., at collapse of the idealised structure), the sum of the positive and negative moments of resistance of the slab must exceed $wL^2/8$, or $17 \times 4^2/8 = 34$ kNm. Within limits (Clause 3.4.2 of CP 110), the designer can choose what proportion of this moment to resist at the supports. Assuming 65%, the required negative moment of resistance is 22 kNm/m. Design in accordance with CP 110 gives 12-mm bars at 200 mm spacing (565 mm²/m). This acts also as top transverse reinforcement for the composite beam (Fig. 3.15). Assuming the factored weight of the steel member to be 2 kN/m, the design load for the composite beam is $17 \times 4 + 2 = 70$ kN/m. Thus

Design moment for beam $= M_c = 70 \times 9^2/8 = 709$ kNm
Design shear for beam $\quad = V_c = 70 \times 4.5 \quad = 315$ kN

3.4　Flexural strength of composite beams

Tests show that many modes of failure are possible in composite beams. The object of research has been to develop design rules for slab reinforcement, shear connectors, and limiting slenderness of steel flanges and webs such that a properly designed beam fails in flexure at a bending moment not less than that predicted by simple plastic theory. These rules are described later. Unless there is a particular reason why the steel member must be slender, as may occur in the design of a long-span composite plate girder, it can be assumed at this stage that simple plastic theory is applicable, whether 'propped' or 'unpropped' construction is used.

The effective breadth of concrete flange must now be determined. Due to shear strain in the plane of the slab, longitudinal bending stress is not constant across a wide thin flange, but varies as sketched in Fig. 3.5. Simple bending theory will give the correct value of the maximum

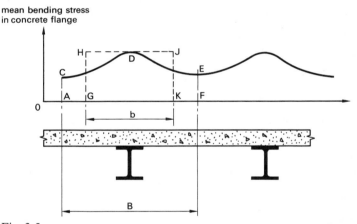

Fig. 3.5

stress (at point D) if the true flange breadth B is replaced by an effective breadth, b, such that the area $GHJK$ equals the area $ACDEF$. Research based on elastic theory has shown that the ratio b/B depends in a complex way on the ratio of B to the span L, the type of loading, the boundary conditions at the supports, and other variables. The results of this work have been boiled down, with some loss of economy, to the simple rule given in CP 110: that the effective breadth shall not extend beyond $L/10$ on either side of the web of the beam.

Little research has been done on effective breadth at flexural failure, so the rules based on elastic theory are widely used at the ultimate limit state. One study of the subject[24] has concluded that for composite beams, it is on the safe side to do this.

The calculation of ultimate moment of resistance is a simple application of the rectangular-stress-block theory used for structural steel and reinforced concrete members. As in CP 110, the concrete is assumed to resist a compressive stress of $0.6 f_{cu}/\gamma_m$, but no tension. The steel is assumed to yield in tension or compression at f_y/γ_m. Introducing the factors γ_m, these stresses are $0.4 f_{cu}$ and $0.87 f_y$. The contribution made by the longitudinal reinforcement in the slab to the flexural strength is small, and can usually be neglected. If account is taken of reinforcement in compression, the stress in it should be limited in accordance with CP 110, and it should be adequately restrained against buckling.

When the plastic neutral axis lies in the slab (depth y_1 in Fig. 3.6(a)), the assumed stress distribution at flexural failure is as in Fig. 3.6(b). If it lies below the slab (depth y_2), then a depth $(y_2 - h_c)$ of the steel member (of area A_{sc}) is at yield in compression; but it is more convenient in calculations to retain the tensile-stress block $0.87 A_s f_y$ and to compensate for this by assuming the compressed region to be at *twice* the yield stress (Fig. 3.6(c)).

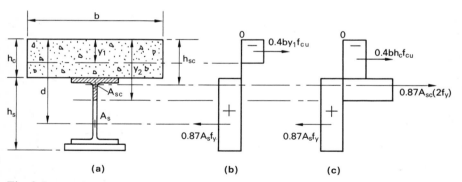

(a)   **(b)**   **(c)**

Fig. 3.6

The following procedure is a convenient way of designing the cross-section of a composite beam to resist a known moment M_c and vertical shear V_c at the ultimate limit state.

1. With guidance from recommended span–depth ratios[32] for composite beams, guess h_s (Fig. 3.6(a)). These ratios are likely to be given in the revised BS 449.

2. Assume the lever arm to be $(h_c + h_s)/2$ and hence estimate the area of steel, A_s, from

$$0.87A_s f_y (h_c + h_s)/2 = M_c \qquad (3.1)$$

3. Check that the available compressive force in the concrete slab, $0.4bh_c f_{cu}$, exceeds $0.87A_s f_y$. If it does not, the plastic neutral axis will be in the steel, which is unusual in beams in buildings, and A_s as given by (3.1) will then be too small.

4. Select a steel section, and check that the shear capacity of its web, $0.87A_w f_y/\sqrt{3}$, exceeds V_c. It is here assumed that the web, of area A_w, carries the whole of the shear, at a stress not exceeding the design yield stress in shear, $f_y/\sqrt{3}\gamma_m$.

5. Calculate the moment of resistance of the section, M_p, as follows and check that it is not less than M_c.

3.4.1   Moment of resistance

If condition 3 above is satisfied, the plastic neutral axis lies in the slab, at depth y_1, found from equilibrium of longitudinal forces:

$$0.4by_1 f_{cu} = 0.87A_s f_y \qquad (3.2)$$

If the centroid of the steel section (which may be asymmetrical, as shown in Fig. 3.6(a)) is at depth d, then by taking moments,

$$M_p = 0.87A_s f_y (d - y_1/2) \qquad (3.3)$$

If the concrete flange is too small to satisfy condition 3, the plastic neutral axis lies in the steel, at depth y_2. The stress blocks are then assumed to be as in Fig. 3.6(c), where A_{sc} is the area of steel at yield in compression. Resolving longitudinally gives

$$0.4bh_c f_{cu} + 1.74A_{sc} f_y = 0.87A_s f_y \qquad (3.4)$$

This gives A_{sc} and hence h_{sc}. Taking moments,

$$M_p = 0.87A_s f_y (d - h_c/2) - 1.74A_{sc} f_y (h_{sc} - h_c/2) \qquad (3.5)$$

3.4.2   Example (continued). Flexural strength of composite beam

The clauses on the serviceability limit state in the draft BS 449[23] are discussed in Section 3.7. For the present design, they give the limiting span–depth ratio as about 15. They are often conservative in particular cases, so a ratio of 16 is first tried here. This corresponds to a beam depth $(h_c + h_s)$ of 0·56 m, and a value $h_s = 0·41$ m. From (3.1),

$$A_s = \frac{709\ 000}{0·87 \times 350 \times 0·28} = 8300 \text{ mm}^2$$

Then $0·87A_s f_y = 2530$ kN, and $0·4bh_c f_{cu} = 3240$ kN, so the neutral axis lies in the slab.

The chosen steel section is a 412 mm × 153 mm × 67 kg/m Universal Beam, for which $A_s = 8530$ mm^2; so both h_s and A_s are similar to the preliminary values above. Other dimensions are given in Fig. 3.7. The shear capacity of the web is

$$\frac{412 \times 9·4 \times 0·87 \times 350}{\sqrt{3} \times 1000} = 680 \text{ kN}$$

which exceeds V_c (315 kN).

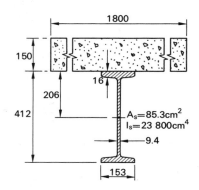

Fig. 3.7

The force in the steel at yield, F_y, is required in several calculations:

$$F_y = 0·87A_s f_y = 2·6 \text{ MN} \tag{3.6}$$

The moment of resistance is now calculated. From (3.2),

$$y_1 = 2 \cdot 6 \times 10^6 / 0 \cdot 4 \times 1800 \times 30 = 120 \text{ mm}$$

and from (3.3) with $d = 356$ mm,

$$M_p = 2 \cdot 6(356 - 60) = 770 \text{ kNm}$$

This exceeds M_c (709 kNm), so the section is adequate.

The plastic modulus of the steel section alone is 1323 cm³, so that its plastic moment of resistance, M_{ps}, is

$$M_{ps} = 0 \cdot 87 \times 1323 \times 350 \times 10^{-3} = 403 \text{ kNm} \qquad (3.7)$$

Thus connection to the concrete increases the available flexural strength by 91%.

3.5   Number and spacing of shear connectors

It was shown in Chapter 2 that slip is inevitable in a composite beam, but that while the beam remains elastic its effects are small. For a given degree of shear connection, loss of interaction due to slip increases as flexural failure is approached, but in a given beam the corresponding loss of strength can be reduced by providing more connectors. There is also a reserve of strength, not considered in simple plastic theory, due to strain-hardening of the steel.

The number of shear connectors required in beams not subjected to repeated loading is based on ultimate-strength behaviour, and is determined by two criteria:

(1) Since shear failure is more sudden and less predictable than flexural failure, there must be sufficient connectors for the beam to fail in flexure, not in shear.
(2) Since partial-interaction calculations are too complex for use in practice, there must be sufficient connectors for the strength of the beam to be not less than that given by simple plastic theory.

An outline of the work by Yam and Chapman[25] that led to the first British recommendations[11] on this subject is now given. At or near ultimate load, there is no simple way of calculating the longitudinal shear at particular points on the steel–concrete interface in a composite beam, but the total shear force to be transmitted is easily found.

Figure 3.8 shows separately the steel and concrete components of one half of a beam of span L that is at flexural failure under a central point load W. If the plastic neutral axis is in the concrete, the longitudinal forces in steel and concrete at midspan are F_y (defined in Section 3.4.2) as shown, and this is the total shear resisted by the connectors

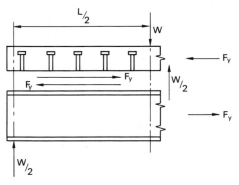

Fig. 3.8

in a half span. If there are N connectors each of push-out strength P_u, the degree of shear connection, r, can be defined by

$$r = NP_u/F_y \tag{3.8}$$

Yam and Chapman devised a method of computing the ultimate moment of resistance M_u of a beam, taking account of the shapes of the stress–strain curves for steel and concrete and the load–slip curve for stud connectors, and analysed many beams like that of Fig. 3.8, using different values of r. The ratio of M_u to the 'simple plastic' moment M_p was plotted against r, as shown in Fig. 3.9. In the lower curve, the effects of strain hardening are neglected. The line AB gives the strength of the steel section alone. Along BC, the shear connection was sufficient to give some composite action, but at such large slips that the studs sheared off before failure of the steel or concrete. Along CD, flexural failure occurred by crushing of the concrete, but M_u was always less than M_p, due to loss of interaction.

When allowance was made for strain hardening, the full-interaction strength was raised from the line $M_u/M_p = 1$ to the line labelled M_{full}, and the curve BCD was raised to EFG. This showed that M_p could be reached when r exceeded about 0·6, but that to ensure flexural failure, rather than shear failure, r had to exceed about 1·1. It was concluded from this and other work that criteria (1) and (2) above were satisfied when $r = 1·25$, but that there was little advantage in using higher values of r. The design strength of a connector, P_d, was therefore defined in CP 117: Part 1 as $0·8P_u$. It is then possible to assume in design that the force F is shared equally between the connectors, so that N, the number required in a half span of the beam shown in Fig. 3.8 is given by

$$F_y = NP_d \qquad (3.9)$$

Putting $P_d = 0·8P_u$ and substituting in (3.8) gives $r = 1·25$. This method is known as *80% design*.

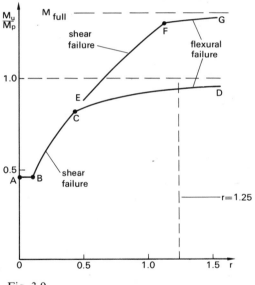

Fig. 3.9

In a similar series of computer tests on beams with distributed loading, 80 per cent design was used, and the connectors were spaced uniformly along some beams and in accordance with the longitudinal shear given by elastic theory ('triangular' spacing) in others. Typical results for the variation of interface shear (q) along the beam just before flexural failure, allowing for strain hardening, are given in Fig. 3.10. At first sight, triangular spacing seems appropriate for distributed loading, but consideration of ultimate-load behaviour has shown uniform spacing to be preferable, for these reasons:

(1) The maximum value of q is lower, and the variation of q along the beam more uniform, giving less severe local stresses in the concrete and simpler detailing of the slab reinforcement in the vicinity of the connectors.
(2) Flexural strength is diminished by loss of interaction only when this occurs at the location of the plastic hinge (at midspan in this example). Uniform spacing gives more connectors in this region.
(3) Uniformly spaced connectors are easier to detail, and fabrication of the steelwork is simpler.
(4) Triangular spacing is in any event not 'correct' when inelastic action and non-uniform loads are considered.

It was concluded that uniform spacing could be allowed for simply supported beams not subjected to heavy concentrated loads. It is now used also in continuous beams (p. 113).

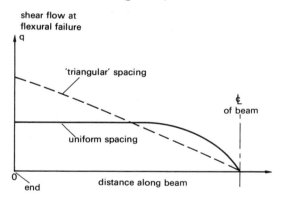

Fig. 3.10

The situation that led to the exclusion of concentrated loads is illustrated in Fig. 3.11. The beam carries its own weight, and concentrated loads at the quarter points, giving the bending-moment diagram ABC. By considering a possible failure of all the shear connectors between the cross-section considered and a free end of the beam, it becomes evident that as one moves away from midspan in a beam with uniform connector spacing, the maximum force F (Fig.

3.8) that can be developed in the slab falls each time a connector is passed, and is only $\frac{1}{2}F_y$ at a quarter point. The bending strength falls less rapidly (for when F is zero there remains the strength of the steel beam), but is as shown by the stepped line ADE. This lies above the parabolic bending-moment curve AGC due to distributed load, but below curve ABC, showing that flexural failure would occur at a quarter-point at a lower load than expected. The remedy is to relate the numbers of connectors in the lengths AB and BC to the relative areas under a diagram of vertical shear, as shown in Fig. 3.11(c). The

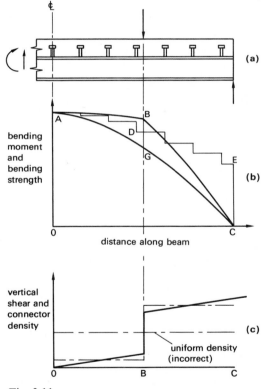

Fig. 3.11

connector spacing between the discontinuities in shear at B and C may be uniform (dashed lines) or in accordance with the vertical shear (continuous line); but it would be unsafe to arrange the same total number of connectors uniformly along the half span (Fig. 3.11(a)).

To control uplift and to avoid too irregular a flow of shear into the concrete, limits are placed on the maximum spacing of connectors along a beam. In current British practice, there are three limits; four times the height of the connector, three times the thickness of the slab, and 0·6 m. There has been no systematic research on this subject, probably because in the past these empirical limits have rarely governed connector spacing. There are now two situations in which relaxation of these limits could lead to economy. These are when partial-interaction design is used (Section 3.9), and in negative-moment regions of continuous bridge beams (Volume 2).

3.5.1   Example (continued). Number and spacing of shear connectors

Subject to the limits on spacing given above, it is usual to select the largest size of connector that will conveniently fit into the slab, for the cost of connectors per unit of shear transferred is higher for the smaller sizes. In the present example, the 19-mm stud 100 mm high is a likely choice. Table 2.1 gives the characteristic strength P_k as 100 kN when the concrete strength is 30 N/mm^2. Multiplying by 0·8, as explained above, and dividing by the partial safety factor (1·15) gives the design load per stud, 69·5 kN. From p. 48, F_y is 2·6 MN, so from (3.9), $N = 2600/69·5 = 38$ studs per half span. These are conveniently arranged in pairs at 0·24 m pitch, which is well below the limiting spacing of 0·4 m.

3.6   Transverse reinforcement in the slab

The formulation of design rules to ensure that the concrete slab has sufficient strength to resist longitudinal shear in the vicinity of the connectors has proved to be difficult, for provision must be made for a great variety of shapes of connector and of haunch, and for the use of precast units; and interaction with the vertical shear V_s and transverse bending moment M_s in the slab must be considered. The mean vertical shear stress on a plane such as BB in Fig. 3.12 is usually so much less than the longitudinal shear stress on that plane that it can be neglected; but as failure is approached, the moment M_s causes cracks in the slab, as shown at C, which may be expected to influence the shear strength of plane BB.

The top transverse reinforcement per unit length of beam, A_t, required to resist the slab moment M_s is found when the slab is designed, and some bottom reinforcement in the slab is likely to extend to the supporting beam, but may not be continuous across it. The problem now is to decide how much more reinforcement is needed to prevent shear failure of the concrete, and where it should be placed.

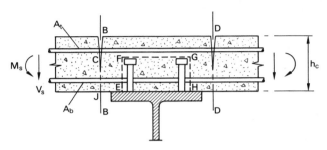

Fig. 3.12

Tests show that shear failure tends to occur on longitudinal planes similar to those shown in Fig. 2.1, but when sufficient shear connectors are provided, the critical planes pass round the connectors or through the slab. In Fig. 3.12, the critical shear surfaces are obviously BB, DD, and $EFGH$. Following the usage established in CP 117, the symbol Q is now used for the design longitudinal shear force at the ultimate limit state on any shear plane through the concrete, of length L_s. For the plane $EFGH$, Q is obviously equal to q, the longitudinal shear per unit length transferred by the connectors; but for other planes, Q is less than q.

In a T-beam, the shear on a plane such as BB is usually taken as $q/2$, but in an L-beam or when the flange of the steel member is wide (Fig. 3.13), the more accurate expressions should be used:

$$Q_{BB} = qb_1/b \quad \text{and} \quad Q_{DD} = qb_2/b \tag{3.10}$$

where b is the effective breadth of the flange.

It is convenient to discuss the results of the relevant research on this subject in terms of v_u, the mean ultimate shear stress on the plane of shear failure. It is given by

$$v_u = Q/L_s \tag{3.11}$$

In 1969 a study was made[26] of all available results of tests on composite beams in which longitudinal shear stresses were high. These included a few failures by shearing or splitting along longitudinal planes. It was found that the design methods of CP 117 Parts 1 and 2 were unduly conservative, and that there was good agreement with the results of tests at the University of Washington on shear transfer in

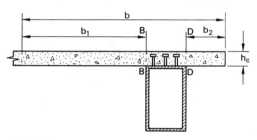

Fig. 3.13

reinforced concrete slabs. A new design method, based on this work, was proposed, and was published in 1971[27] in a form relevant to the contemporary draft code of practice.[15] The method to be described here is that now proposed for use in the codes of practice for composite structures. It is a development of the 1971 method, based mainly on further work on shear transfer in slabs.[28]

3.6.1   Design method

In 1972, Mattock and Hawkins showed that a safe estimate of the maximum shear that could be transferred across a pre-existing crack in a reinforced concrete slab was given by

$$v_u = 200 \text{ lb/in}^2 + 0\cdot8(pf_r + f_n) \tag{3.12}$$

where p is the area of reinforcement per unit area of shear plane, f_r is its yield strength, and f_n is the mean externally applied direct compressive stress acting normal to the shear plane. They also placed limits on v_u and $(pf_r + f_n)$. These are discussed in Section 3.6.3.

The application of (3.12) to the shear plane $EFGH$ (of length L_s) in Fig. 3.12 in the presence of a transverse negative moment M_s per unit length of beam is now considered. If v_u in the equation is taken as the design shear at the ultimate limit state, it follows that the material strengths given or implied in (3.12) should also be assumed to be design values. In codes of practice, design equations are normally given in terms of characteristic strengths of materials, so in (3.12) f_r is now replaced by $f_{ry}/1.15$, and 200 lb/in^2 (which represents the shear strength of the concrete in absence of transverse compressive stress) by $(200/1.5)$ lb/in^2, which is 0.9 N/mm^2. Multiplying by L_s and using (3.11), eq. (3.12) then becomes

$$Q = 0.9sL_s + 0.7A_e f_{ry} + 0.8f_n L_s \tag{3.13}$$

where s is a constant stress of 1 N/mm^2, to be expressed in units consistent with those used for the other quantities, and A_e is the area of reinforcement per unit length of beam that crosses the shear plane and is fully anchored on both sides of it. The function of s is simply to make the equation independent of units.

The influence of the transverse bending moment in the slab is allowed for by calculating the transverse compressive force F_c per unit length of beam that acts on the surfaces EF and GH of the shear plane. This is easily done when the slab is designed. For example; if the connectors extend through at least half the thickness of the slab and the top reinforcement is fully stressed, then the compression in the concrete is equal to the tension in the reinforcement when at its design yield stress:

$$F_c = A_t f_{ry}/\gamma_m \tag{3.14}$$

where $\gamma_m = 1.15$. The total normal force on the shear plane is $2F_c$, so for design purposes (3.13) becomes

$$Q \ngtr 0.9sL_s + 0.7A_e f_{ry} + 1.6F_c \tag{3.15}$$

This expression is used to calculate the amount of bottom transverse reinforcement, putting $A_e = 2A_b$ for plane $EFGH$. When the loading is uniformly distributed over the slab, (3.15) can usually be satisfied with $A_b = 0$.

Mattock and Hawkins showed (3.12) to be valid whether f_n was tensile or compressive. It follows that F_c in (3.15) can be taken as the resultant force normal to the shear plane. If membrane effects are neglected, this force is always zero on a plane through the whole thickness of a slab. The design equation for planes such as *BB* in Fig. 3.12 is therefore

$$Q \not> 0.9sL_s + 0.7(A_t + A_b)f_{ry} \tag{3.16}$$

No limits need be placed on the ratio of A_t to A_b, and the top transverse reinforcement can be assumed to be effective in resisting both transverse bending and longitudinal shear. This is in accordance with extensive data from tests on composite beams. It is explained partly by dowel action and aggregate interlock across crack *BC*; but a more significant factor is that at ultimate load, the tension $A_t f_{rd}$ in the top reinforcement is associated with an equal transverse compressive force in the concrete on plane *CJ*, which increases the shear strength of that plane by an amount proportional to $A_t f_{rd}$.

3.6.2   Longitudinal shear in haunched beams

In Section 2.5, limits are proposed for the dimensions of haunches for which the strengths of shear connectors can be assumed to be as for unhaunched beams. A shear plane through such a haunch (*EFGH* in Fig. 3.14) is now considered.

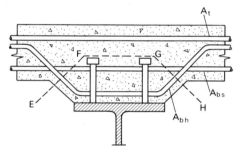

Fig. 3.14

When the reinforcement is as shown, it may be difficult to decide whether the bottom reinforcement in the slab, of area A_{bs}, can be included in the 'effective' reinforcement A_e. The criterion is whether these bars are low enough to prevent a failure involving vertical separation of the slab from the beam. It is recommended[23] that they be included if they are placed at a clear distance of not less than 30 mm below the surface of each shear connector that resists uplift forces.

This proposal is based on Clause 11.1.5(2) of CP 117: Part 2, which requires bottom reinforcement in a haunch to be at least 2 inches below the top of the connectors.

Another question concerning haunches is the value of F_c to be used in (3.15). The highest local stresses in the concrete occur near the bottom of the shear connectors, where the influence of transverse compression is small; so it would be prudent to take F_c as zero. This is also the reason for another proposal: that in haunches at least half of the reinforcement required by (3.15) should pass within 50 mm of the top surface of the steel member, and so contribute to the strength of the highly stressed region.

There is a need for more tests on haunched beams, for these are likely to show that the proposed rules are conservative for shallow haunches.

Research on haunches that do not comply with the limits given in Section 2.5 has shown[29, 30] that the reinforcement should be similar to that required in the webs of reinforced concrete T-beams.

3.6.3   Limits to the applicability of the design method

Two limits were given[28] to the applicability of eq. (3.12): that v_u should not exceed $0.3f_c$, where f_c is the cylinder strength of the concrete, and that $(pf_{ry} + f_n)$ should not be less than 200 lb/in^2.

Putting $f_c = 0.8f_{cu}$, and dividing by γ_m for concrete, as before, and multiplying by L_s, the first limit becomes $Q \not> 0.16L_s f_{cu}$. The corresponding limits given in CP 117 are proportional to $(f_{cu})^{1/2}$, on the assumption that the governing factor is the tensile strength of concrete. Tests on composite beams have shown[26] that the limit on v_u should lie above $1.04(sf_{cu})^{1/2}$; but there is not enough evidence to decide between f_{cu} and $(f_{cu})^{1/2}$ as the relevant variable, so f_{cu} is now chosen for simplicity. After allowing for γ_m for concrete, the limit $1.04(sf_{cu})^{1/2}$ gives values ranging from $0.12f_{cu}$ to $0.19f_{cu}$, and so agrees fairly well with Reference 28. The limit now proposed, based on the evidence available, is

$$\left. \begin{array}{l} Q \not> 0.15L_s f_{cu} \\ f_{cu} \not> 45 \text{ N/mm}^2 \end{array} \right\} \qquad (3.17)$$

with the restriction

The rule $pf_{ry} + f_n \not< 200$ lb/in^2 is based partly on tests on corbels that are not relevant to composite beams. In effect, it gives a minimum amount of transverse reinforcement. CP 110 provides a better guide. Clause 3.11.4.2 requires that the top transverse reinforcement in a flanged beam shall not be less than 0·3%, and shall extend across the full effective width of the flange. The type of steel is not specified. If mild steel bars are used, the rule becomes $pf_{ry} \not< 0.75$ N/mm^2. It follows from the preceding discussion that it is unnecessary to require that the minimum reinforcement be placed near the top of the slab. It is suggested that the 0·75 be rounded up to 0·8 N/mm^2, and the rule given in the form

$$A_b + A_t \not< 0.8sh_c/f_{ry} \tag{3.18}$$

where h_c is the thickness of the slab assumed to be effective as the flange of the composite beam.

3.6.4   Detailing of transverse reinforcement

Clause 3.11.8.2 of CP 110, 'Maximum distance between bars in tension' is applicable to top transverse reinforcement; but no rule is given for bars in compression or shear. Clause 10.5 of CP 117: Part 2 limits the spacing of bottom transverse reinforcement to three times the projection of the shear connectors above the bars. The intention is to ensure that local shear failures on planes such as $EFGH$ (Fig. 3.12) do not occur between widely spaced bars. In the draft Bridge Code, the 'three' in this empirical rule has been increased to 'four', but an overall limit on bar spacing of 600 mm has been added. In this amended form it is recommended for use with the design method described above, but with a further relaxation. Any rule of this kind is restrictive when (3.15) is applicable and the amount of bottom transverse reinforcement required is found to be small. It also causes problems with the detailing of precast floor slabs, as is evident in Section 3.6.8. It is proposed that when (3.15) is applicable and

$$0.7A_e f_{ry} < 1.6F_c \tag{3.19}$$

the spacing of the bottom bars should not exceed eight times the projection of the connectors above them, nor exceed 600 mm.

3.6.5 Comparison with other design methods

The method described above is superficially similar to those given in CP 117: Parts 1 and 2. An account of the origin of the earlier methods is available.[27] The new method in fact represents a substantial relaxation in requirements for transverse reinforcement, and will in practice often enable bottom reinforcement to be omitted altogether. Detailed comparisons are made difficult by the change in design philosophy and the fact that in CP 117, flexural reinforcement was not allowed to contribute to shear strength.

This relaxation is based on a conservative interpretation of the test data that have become available since CP 117 was written; and is necessary, even though it may be thought to be complicated, because Britain is almost alone amongst countries that use composite structures in having any rules at all, apart from those in force for reinforced concrete T-beams. Perhaps because the CP 117 rules make it very difficult to detail composite beams with precast floor slabs, other members of the European Economic Community have been more successful in developing and selling such structures. In some of these, the transverse reinforcement does not comply with the new method; yet the structures have been shown by full-scale testing to be satisfactory. The reason is probably that in a real structure, the in-plane stiffness of the slab and the compressive membrane forces that develop in it tend to prevent splitting or shearing failures along rows of connectors, whereas the design method given above is based mainly on research on isolated slabs and on beams with narrow flanges. Therefore it will often be too conservative; but it is still useful because it is less restrictive than CP 117. A general design method that takes account of membrane forces has yet to be devised.

3.6.6 Lightweight-aggregate concrete

As explained in the Handbook to CP 110, the shear strength of lightweight-aggregate concrete is less than that of normal-density concrete with the same cube strength. The reduction is assumed in CP 110 to be 20%, and is equally applicable in composite beams. For shear planes through concrete made with lightweight aggregate, the term $0.9sL_s$ in expressions (3.13), (3.15), and (3.16) should therefore be replaced by $0.7sL_s$; and the term $0.15L_sf_{cu}$ in (3.17) by $0.12L_sf_{cu}$.

3.6.7 Example (continued). Transverse reinforcement in the slab

From p. 54, studs with a design strength of 69·5 kN are placed in pairs at 0·24 m pitch, so the shear force per unit length on plane $EFGH$ in Fig. 3.15 is

$$Q = 69\cdot5/0\cdot12 = 580 \text{ N/mm}$$

In designing the slab it was found that 515 mm²/m of top transverse reinforcement was required for a distributed load of 17 kN/m², and 565 mm²/m was provided (p. 44).

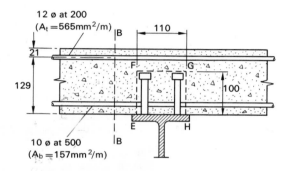

12 ø at 200
$(A_t = 565 \text{mm}^2/\text{m})$

10 ø at 500
$(A_b = 157 \text{mm}^2/\text{m})$

Fig. 3.15

For safe design, the force F_c should be estimated conservatively. Here, let us suppose that the characteristic imposed load on the slab (7·5 kN/m²) includes an allowance of 1·5 kN/m² for partitions. The whole of the partition load might in practice be placed directly above the beam, and so not cause transverse bending of the slab. The effective design load for transverse bending is therefore reduced by

$$1\cdot5 \times 1\cdot6 = 2\cdot4 \text{ kN/m}^2$$

so the amount of top reinforcement that can be assumed to be at yield in tension is

$$515 \times (17 - 2\cdot4)/17 = 441 \text{ mm}^2/\text{m}$$

and from (3.14),

$$F_c = 0\cdot441 \times (410/1\cdot15) = 157 \text{ N/mm} \tag{3.20}$$

The length of plane $EFGH$ is 310 mm, and for this plane $A_e = 2A_b$. Using (3.15) to find A_b:

$$580 = 0.9 \times 310 + 1.4 \times 410 A_b + 1.6 \times 157$$

whence

$$A_b = 87 \text{ mm}^2/\text{m} \tag{3.21}$$

and $1.6F_c > 0.7 A_e f_{ry}$. If 8-mm bars with 15-mm cover are used as bottom reinforcement, the connectors project 77 mm above these bars, so their spacing may not exceed $8 \times 77 = 616$ mm, nor 600 mm, which governs. At this stage it is prudent to consider the requirements of CP 110 in respect of curtailment and anchorage of reinforcement in slabs. Clause 3.11.7.3 gives the 'simplified rule' that 50% of the tension reinforcement at midspan should extend into the support. In designing the slab, the moment at midspan was assumed to be 12 kNm/m (p. 44). The corresponding reinforcement is 273 mm²/m, so that 137 mm²/m must cross the shear plane FE, but need not be fully anchored beyond it; so requirement (3.21) above, which relates to fully anchored bars, is not wholly superseded. In practice it would be simplest to provide 137 mm²/m of fully anchored bars; so 10-mm bars at 0.25 m spacing might be provided at midspan, with alternate bars continuous over the steel beam, giving $A_b = 157$ mm²/m in this region (Fig. 3.15).

The total area of top and bottom reinforcement is then $565 + 157 = 722$ mm²/m, and it is evident that the expressions (3.16) for shear plane BB and (3.18) for minimum reinforcement are satisfied. The value of $Q/L_s f_{cu}$ for plane $EFGH$ is 0.06; well below the limit set by (3.17).

3.6.8   Use of precast concrete slabs

Steel frames tend to be used in buildings when the time available for construction is short. It may then be economical to use prefabricated flooring. Corrugated metal decking (Section 3.10) is widely used in North America, but in Europe precast concrete units are often found to be cheaper. The following example illustrates the problems of detailing that arise when wide precast units are used in composite beams. No attempt is made to optimise the design of the units, for many types are in use, and an account of them would be outside the scope of this book.

Clause 5.2.4.3 of CP 110 states that whenever possible, precast units should have a bearing of at least 75 mm when supported on steel beams. To allow the placing of concrete between the units (Section 3.3), the gap between their ends should be at least 30 mm, so the detail shown in Fig. 3.4 is not possible when the steel flange is only 153 mm wide, as in the current example. The section previously chosen (Fig. 3.7) is therefore replaced by a 454 mm by 190 mm by 67 kg/m Universal Beam, which has the same cross-sectional area. For comparison with Section 3.6.7, the design shear force per unit length will be taken as 580 N/mm, as before.

It is assumed that the precast units span between the steel beams at 4·0-m centres, and are 75 mm thick and of rectangular cross-section. They are to be erected on props, which are to be left in place until 75 mm of in-situ topping has hardened, so the effective thickness of the slab for all loads is 150 mm, as before.

The bottom reinforcement at midspan may be determined by handling stresses. We assume that a unit is lifted by its ends when the cube strength is not less than 30 N/mm^2, and allow for a vertical acceleration of 0·3g. Design to CP 110 leads to the provision of 8-mm bars at 0·15 m spacing (336 mm^2/m) at midspan, with a bottom cover of 15 mm. In the completed structure, the whole of this reinforcement can be assumed to be effective in the 150-mm slab at the ultimate limit state, and it provides a flexural strength of 13·7 kNm/m. It was found in Section 3.3.1 that $wL^2/8$ for a span of the slab is 34 kNm/m. The midspan reinforcement of 336 mm^2/m provided 40% of $wL^2/8$, which is sufficient for an internal span, and leaves 20·3 kNm/m to be provided over the supports. Design to CP 110 shows that 12-mm bars at 0·24 m spacing (473 mm^2/m) are just sufficient for this purpose. These are placed in the in-situ concrete, as shown in Fig. 3.16.

The detailing near the shear connectors is now considered. It can be deduced from the design of the in-situ slab (Section 3.6.7) that some bottom transverse reinforcement will be required. Even if the bearing of the precast units on the steel flange is 60 mm, rather than the 75 mm recommended in CP 110, the gap between the ends of the units is only 70 mm, which is not enough space in which to provide shear connectors and to anchor the bottom reinforcement. It follows that the ends of the slabs must be cut back at each connector. The connector spacing should therefore be as wide as possible. The static strength of 22-mm diameter headed studs 100 mm high in Grade 30 concrete is given in the draft Bridge Code[15] as 126 kN, 26% above that of the 19-mm studs used previously in pairs at 0·24 m spacing. Pairs of the larger studs at a spacing of $0·24 \times 1·26 = 0·302$ m would be sufficient. They are therefore provided at 0·3-m centres, in place of the 19-mm studs.

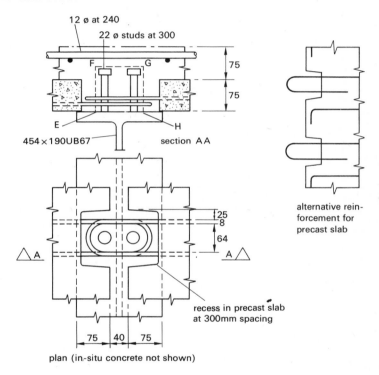

Fig. 3.16

The next step is to find out how much bottom transverse reinforcement must be fully anchored within the potential shear plane of type *EFGH* (Fig. 3.16). In calculating the force F_c, it would be prudent to neglect any transverse compression in the in-situ concrete between the ends of the precast units due to the weight of the units themselves. This and the omission of the partition loading (as before) together reduce the effective area of top transverse reinforcement from 473 mm^2/m to 337 mm^2/m, so from (3.14),

$$F_c = 0.337 \times (410/1.15) = 120 \text{ N/mm} \tag{3.22}$$

To provide bearing for the units, the connectors in each pair are more closely spaced than before, so the length FG (Fig. 3.16) is reduced from 110 mm to 94 mm, giving $L_s = 294$ mm. Putting $A_e = 2A_b$, as before, in (3.15) gives

$$580 = 0.9 \times 294 + 1.4 \times 410A_b + 1.6 \times 120$$

whence

$$A_b = 216 \text{ mm}^2/\text{m} \tag{3.23}$$

The midspan bottom reinforcement is 8-mm bars at 0.15-m spacing (336 mm^2/m), and (3.23) shows that two-thirds of this should be anchored in the gap between the units. As the connector spacing is twice the spacing of these bars, it is almost as simple to anchor all of it, so leaving a margin of strength to allow for any holes in the slab. One way in which this could be done is shown in Fig. 3.16. This detail is not ideal, for the bottom bars now have irregular spacing and, more important, the projections of the slabs between pairs of connectors would require further reinforcement, for they will transfer most of the vertical shear to the steel beam. The inset in Fig. 3.16 shows a way in which both these problems could be solved, still using 8-mm bars at 0.15-m spacing.

In this example, the proposed design method leads to half the quantity of bottom transverse reinforcement that would be required by CP 117: Part 1; but it still involves erection of the slabs in a predetermined sequence. This is almost inevitable when the slabs are too wide for the bottom reinforcement to be placed in the joints between them, as shown in Fig. 3.4.

3.7 Stresses and deflections in service

When a composite beam has been designed by ultimate-strength methods, checks must be made to ensure that at service loads, its deflection is not excessive and the stresses are within the elastic range.

This can be done by elastic analysis, using the transformed-section method and assuming full interaction.

Where no relevant ultimate-strength theory exists or when failure may occur by elastic instability, as in composite bridge beams with slender webs and flanges, design for the ultimate limit state must also be based on elastic theory, and then stresses and deflections at working load are less likely to influence design.

The following examples show that elastic analysis of composite beams takes much longer than ultimate-strength analysis, and is more complex than that for reinforced concrete beams. If accurate results are required, account must be taken of the method of construction, of shrinkage of concrete, and of the effects of creep. This involves the separate consideration of three types of loading: that carried by the steel beam; and the long-term and short-term loads carried by the composite beam.

These problems have led to the development of other methods of meeting serviceability criteria. Deflections can be controlled by designing to limiting span–depth ratios. Values for reinforced concrete structures are given in CP 110, and it has been found that in buildings no limits need be placed on stresses in reinforcement or in concrete at the serviceability limit state, for the other criteria of strength and limiting deflection are sufficient.

In concrete bridges, it is more rational to limit stress in reinforcement at the serviceability limit state than to impose an arbitrary limit on deflections. In the 1973 draft of Part 4 of the Unified Bridge Code, 'The design of concrete bridges', this limit is set at $0.8f_{ry}$.

In composite structures, there are two reasons why yield of structural steel under working loads may be more of a problem than yield of reinforcement: the effects of unpropped construction, and of residual stresses in rolled steel sections. It seems likely that the Bridge Code will adopt the same limiting stress (80% of yield) for structural steel in composite members as has been proposed for reinforcement.

In buildings where fatigue is not a problem, the consequences of yield of steel at working load are less important, so a higher limit can be used. If the proposals described on p. 73 for the control of deflections by means of limiting span–depth ratios are found to be satisfactory, it may be possible to raise the limiting stress in steel in midspan regions at the serviceability limit state from the present

$0.9f_y$ (given in CP 117: Part 1) to $1.0f_y$. This limit is unlikely ever to govern the design of simply supported beams for which propped construction is used. It is not yet clear whether the proposed span–depth ratios for unpropped construction and for continuous beams will enable designers to omit the calculation of bending stresses in the structural steel member in such beams, so the relevant methods are given below and in Chapter 4.

3.7.1   Formulae for full-interaction elastic analysis of cross-sections

This method of analysis differs from that used for reinforced concrete beams only in that allowance is made for the flexural rigidity of the steel section. It is assumed that plane sections remain plane, and the tensile strength of concrete is neglected. For generality, the steel section is assumed to be asymmetrical, Fig. 3.17, with cross-sectional area A_s, second moment of area I_s, and centroid distance d below the top of the concrete slab, which is of uniform thickness h_c.

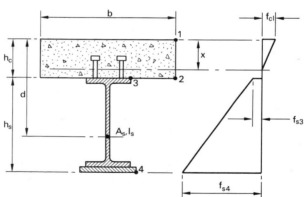

Fig. 3.17

The modular ratio m may have two values. For imposed (short-term) load,

$$m = m_1 = E_s/E_c \tag{3.24}$$

For dead (sustained) load, with a creep factor k_c, the effective modulus of concrete is $k_c E_c$, so that

$$m = m_2 = E_s/k_c E_c \tag{3.25}$$

When concrete in tension is neglected, the formulae for neutral-axis depth x and second moment of area of the composite section, I, depend on the sign of $(h_c - x)$. If

$$mA_s(d - h_c) < \tfrac{1}{2}bh_c^2 \qquad (3.26)$$

then x is less than h_c and is given by the usual 'first moment of area' equation,

$$mA_s(d - x) = \tfrac{1}{2}bx^2 \qquad (3.27)$$

If (3.26) is not satisfied, $x \geqslant h_c$, and then

$$mA_s(d - x) = bh_c(x - h_c/2) \qquad (3.28)$$

The second moment of area is given in 'steel' units by the usual equations, using the parallel-axis theorem. For $x < h_c$,

$$I = bx^3/3m + I_s + A_s(d - x)^2 \qquad (3.29)$$

For $x \geqslant h_c$,

$$I = \frac{bh_c^3}{12m} + \frac{bh_c}{m}(x - h_c/2)^2 + I_s + A_s(d - x)^2 \qquad (3.30)$$

In the analysis of continuous structures, it is sometimes convenient to use stiffnesses of members based on the uncracked concrete section; that is, assuming concrete to be equally stiff in tension and compression. The neutral-axis depth and second moment of area are then as given by (3.28) and (3.30) above, for all values of the ratio x/h_c. In most composite beams, the neutral axis at midspan is close to the steel–concrete interface, and the difference between the 'cracked' and 'uncracked' values of I is negligible.

Expressions are needed for the stresses due to a positive bending moment M in concrete at levels 1 and 2 (f_{c1} and f_{c2}), and in steel at levels 3 and 4 (f_{s3} and f_{s4}). By the elementary theory of bending, with tensile stress positive,

$$f_{c1} = -Mx/mI \qquad (3.31)$$

If $x \geqslant h_c$,

$$f_{c2} = -M(x - h_c)/mI \qquad (3.32)$$

For steel,

$$f_{s3} = M(h_c - x)/I \qquad \text{for all } x, \tag{3.33}$$

and

$$f_{s4} = M(h_c + h_s - x)/I \tag{3.34}$$

3.7.2   Example (continued). Stresses in service

From Fig. 3.7 and Section 3.2, the symbols used above have these values:

$b = 1\!\cdot\!8$ m	$E_s = 200$ kN/mm²	$A_s = 85\!\cdot\!3$ cm²
$h_c = 0\!\cdot\!15$ m	$E_c = 26\!\cdot\!7$ kN/mm²	$I_s = 23\ 800$ cm⁴
$h_s = 0\!\cdot\!412$ m	$k_c = 0\!\cdot\!5$	$d = 0\!\cdot\!356$ m

From (3.24), $m_1 = 7\!\cdot\!5$ for imposed load.
From (3.26), $mA_s(d - h_c) = 13\ 200$ cm³, $\frac{1}{2}bh_c^2 = 20\ 250$ cm³, so $x < h_c$.
From (3.27), $x_1 = 12\!\cdot\!7$ cm.
From (3.29), $I_1 = 84\ 900$ cm⁴.
For dead load, $m_2 = 15$ from (3.25).
From (3.26), $mA_s(d - h_c) = 26\ 400$ cm³, which exceeds $\frac{1}{2}bh_c^2$, so $x > h_c$.
From (3.28), $x_2 = 16\!\cdot\!5$ cm. Thus the increase in m to allow for creep causes a significant change in the neutral-axis depth. Here it is increased by 30%.
From (3.30), $I_2 = 72\ 870$ cm⁴.

The design loads for the serviceability limit state are now found from data given in Section 3.3.1, assuming that the steel member weighs 0·6 kN/m. They are

dead, $4 \times 3\!\cdot\!6 + 0\!\cdot\!6 = 15$ kN/m,
$$\text{whence } M_d = 15 \times 9^2/8 = 152 \text{ kNm};$$
imposed, $4 \times 7\!\cdot\!5 = 30$ kN/m,
$$\text{whence } M_i = 30 \times 9^2/8 = 304 \text{ kNm}.$$
Similarly,

$$V_d = 15 \times 4\!\cdot\!5 = 67\!\cdot\!5 \text{ kN} \qquad V_i = 30 \times 4\!\cdot\!5 = 135 \text{ kN}$$

The longitudinal bending stresses at midspan are now calculated for both 'propped' and 'unpropped' construction, using eqs. (3.31)–(3.34). The symbols with suffix 1 correspond to imposed load, and those with suffix 2 to dead load.

(1) If all load is carried by the composite section,

$$f_{c1} = -\frac{M_i x_1}{m_1 I_1} - \frac{M_d x_2}{m_2 I_2} = -6 \cdot 1 - 2 \cdot 3 = 8 \cdot 4 \ \text{N/mm}^2$$

$$f_{c2} = 0 - \frac{M_d(x_2 - h_c)}{m_2 I_2} = 0 - 0 \cdot 21 = -0 \cdot 21 \ \text{N/mm}^2$$

$$f_{s3} = \frac{M_i(h_c - x_1)}{I_1} + \frac{M_d(h_c - x_2)}{I_2} = +8 \cdot 1 - 3 \cdot 1 = +5 \cdot 0 \ \text{N/mm}^2$$

$$f_{s4} = \frac{M_i(h_c + h_s - x_1)}{I_1} + \frac{M_d(h_c + h_s - x_2)}{I_2}$$
$$= +155 + 83 = +238 \ \text{N/mm}^2$$

The reader may wonder why f_{c2} is compressive and f_{s3} is tensile, when both relate to points at the same level in the member. The reason is that in adding the two stresses that make up f_{c2} we have assumed that concrete cracked in tension (due to imposed load) can resist compression (due to dead load); whereas in fact the concrete probably would not crack at such a low tensile stress. This serves to illustrate the fact that there is little point in calculating flexural stresses in concrete by the elastic theory. In limit-state design, there is no need to do so.

(2) If the dead load is carried by the steel section alone. The steel section is symmetrical, so the dead-load stresses are

$$f_{s3}, f_{s4} = \mp M_d h_s / 2 I_s = \mp 131 \ \text{N/mm}^2$$

The imposed-load stresses in steel and concrete are as found in (1) above, giving totals

$$f_{c1} = -6 \cdot 1 \qquad f_{c2} = 0 \qquad f_{s3} = -123 \cdot 5 \qquad f_{s4} = +287 \ \text{N/mm}^2$$

There are in addition some small stresses due to the fact that the weight of the formwork is imposed on the steel section and subsequently removed from the composite section. These are neglected.

The stress distributions are plotted on Fig. 3.18, which shows that use of the steel member to carry the wet concrete leads to a substantial increase in stress f_{s3}, which is unimportant here, but could influence the design of an asymmetrical section. There is also a smaller increase in f_{s4}, which may be important if design is on an elastic basis or if it causes the yield stress of the steel to be exceeded. Here the stress is $0.82f_y$, so the stresses due to service loads do not influence the design of the member.

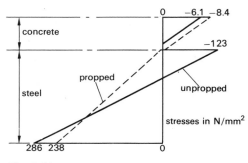

Fig. 3.18

3.7.3   Limiting deflections, and the use of span–depth ratios

It is almost as much an art as a science to ensure at the design stage that the deflection of a slender beam will not be excessive in service. It is possible to allow in calculations for some of the factors that influence deflection, such as creep and shrinkage of concrete; but there are others that cannot be quantified. In developing the limiting span–depth ratios for CP 110, Beeby[31] identified nine reasons why deflections of reinforced concrete beams in service were usually less than those calculated by the designers, and increased his theoretical span–depth ratios by 36% to allow for them. Many of the reasons apply equally to composite beams, the most significant of them being the variations in the elasticity, shrinkage and creep properties of the concrete, the stiffening effect of finishes, and restraint and partial fixity at the supports.

The other problem is the difficulty of defining when a deflection becomes 'excessive'. In practice, complaints often arise from the cracking of plaster on partition walls, which can occur when the deflection of the supporting beam is as low as span/800.[31] For partitions and in-fill panels generally, the relevant deflection is that which takes place after their construction. This can exceed that due to the finishes and imposed-load, for dead-load creep deflections continue to increase for several years after construction. It is good practice to provide partitions with appropriate joints and clearances; it has been proposed[23] that when this is not done the relevant deflection should not exceed span/350, or 20 mm (which governs for spans exceeding 7 m). Where appearance is the only criterion, it is recommended that the full-load deflection, including the effects of creep and shrinkage, of a suspended span below the level of its supports should not exceed span/250; and for roof beams constructed to a fall, an even greater deflection may be acceptable. The difficulty of assessing the accuracy and significance of a calculated deflection is such that simplified methods of calculation are justified. Such a method is given for reinforced concrete beams in Appendix A of CP 110, and a similar method for composite beams is likely to be included in the new BS 449.

The use of span–depth ratios to control deflections is simpler still, and an appropriate method has been developed.[32] Limiting ratios of span to overall depth have been derived for simply supported propped composite beams such that the deflection due to distributed load should not exceed span/250. These are given in terms of the density of the concrete, the ratio of depth of slab to depth of steel member, the ratio of area of concrete to area of steel, and the grade of the structural steel. There are also modification factors to allow for various degrees of end fixity; to ensure lower deflection ratios in long spans, and for beams where the flexural strength provided exceeds that required at the ultimate limit state.

No calculations of stress at the serviceability limit state are needed unless unpropped construction is used. Then the limiting span–depth ratio can be calculated as follows. First the midspan extreme-fibre tensile stress f_{s4} in the steel beam due to the load carried by it alone is found. It can usually be assumed that this load (its own weight and that of the wet concrete) is uniformly distributed. Then the limiting span–depth ratio (R_s) for the steel beam with this maximum stress is found. For distributed loading and a span–deflection ratio of 250, it is

$$R_s = E_s/52f_{s4} \qquad (3.35)$$

For a beam of overall depth h with a steel member of depth h_s, the span–depth ratio for unpropped construction is given by

$$R_u = \frac{R_p}{1 - \lambda + (hR_p/h_sR_s)} \tag{3.36}$$

where λ is the ratio of load carried by the steel beam to the total load at the serviceability limit state, and R_p is the span–depth ratio for the beam if constructed unpropped. An example of the use of this result is given below.

Many conservative assumptions have to be made when deriving limiting span–depth ratios, so that if the proposed ratio for a particular beam slightly exceeds the limiting value, it is likely that calculation will show that the deflection of the beam should be satisfactory. The method is as for steel beams, but using appropriate values of E and I, and is illustrated below. An approximate method of allowing for the additional deflection due to shrinkage is given in Section 3.8.

3.7.4   Example (continued). Deflection in service

The midspan deflection, δ, of a simply supported beam with distributed load and midspan bending moment M is given by

$$\delta = \frac{5wL^4}{384EI} = \frac{5ML^2}{48EI}$$

Working in 'steel' units, with $E_s = 200$ kN/mm^2 and $L = 9$ m,

$$\delta = \frac{5 \times 81 \times 10^5 M}{48 \times 200I} = 4220M/I \text{ mm}$$

when M is in kNm and I is in cm^4. The appropriate values of M and I are given on p. 70.

The imposed-load deflection is thus $4220 \times 304/84\,900 = 15$ mm, which is small (span/600), as is usual in composite beams. For propped construction, the dead-load deflection is $4220 \times 152/72\,870 = 9$ mm, which is also small; but if unpropped construction is used, the relevant I is that of the steel section alone, 23 800 cm^4 in place of 72 870 cm^4, and the dead-load deflection is 27 mm. It is found in Section 3.8 that the shrinkage deflection of this beam is 7 mm, giving the following deflections:

propped construction; $15 + 9 + 7 = 31$ mm (span/290)
unpropped construction; $15 + 27 + 7 = 49$ mm (span/184).

For propped construction, an estimate of the deflection relevant to partitions can be made by assuming that half the dead-load deflection occurs after they have been constructed; this gives a total of 26·5 mm, or span/340, which may or may not be acceptable, depending on the design of the partitions. For unpropped construction, the dead-load deflection of 27 mm includes no creep, so the relevant deflection is $15 + 7 = 22$ mm, which should be satisfactory. The total deflection of 49 mm is likely to be unacceptable, but can be reduced to say 19 mm by pre-cambering the steel member by about 30 mm.

The proposed span–depth ratios[32] do not take account of pre-cambering, so from these results it is to be expected that they should indicate that the beam is near the borderline of acceptability if propped, and is too slender if unpropped. For a propped beam, the calculation is simple. For each grade of steel, the span–depth ratio R_p is given in terms of the ratios of depth of concrete to depth of steel (here 0·364) and area of concrete flange to area of steel beam (here 31·6). For these values and Grade 50 steel, $R_p = 15$. This is below the actual ratio for this beam, which is 9/0·562, or 16·0. For an unpropped beam, the method outlined above is followed. The extreme-fibre stress in the steel beam due to dead load is, from p. 71,

$$f_{s4} = + 131 \text{ N/mm}^2$$

From (3.35),

$$R_s = 29·2$$

From p. 70,

$$\lambda = 152/456 = 0·33 \qquad \text{and} \qquad h = 0·562 \text{ m}$$
$$h_s = 0·412 \text{ m} \qquad\qquad R_p = 15$$

so from (3.36),

$$R_u = \frac{15}{0·67 + (0·562 \times 16·4/0·412 \times 29·2)} = 10·5$$

This shows clearly that the unpropped beam (with $R_u = 16 \cdot 0$) would have a total deflection exceeding span/250, as was found above.

3.8   Effects of shrinkage of concrete and of temperature

In the fairly dry environment of a building, an unrestrained concrete slab could be expected to shrink by $0 \cdot 03\%$ of its length (3 mm in 10 m) or more. In a composite beam, the slab is restrained by the steel member, which exerts a tensile force on it, through the shear connectors near the free ends of the beam, so its apparent shrinkage is less than the 'free' shrinkage. The loads on the shear connectors act in the opposite direction to those due to dead and imposed load, and so can be neglected in design.

The stresses due to shrinkage develop slowly, and so are reduced by creep of the concrete, but the increase they cause in the deflection of a composite beam may be significant. An approximate and usually conservative rule of thumb for estimating this deflection in a simply supported beam is to take it as equal to the long-term deflection due to the weight of the concrete slab acting on the *composite* member.

In the beam studied in the Examples in this Chapter, this rule gives an additional deflection of 9 mm, whereas the calculated long-term deflection due to a shrinkage of $0 \cdot 03\%$ (with a creep coefficient $k_c = 0 \cdot 5$) is 7 mm.

In beams for buildings, it can usually be assumed that tabulated span–depth ratios are sufficiently conservative to allow for shrinkage deflections; but the designer should be alert for situations where the problem may be unusually severe (e.g., thick slabs on small steel beams, electrically heated floors, and concrete mixes with high 'free shrinkage').

Composite beams also deflect when the slab is colder than the steel member. Such differential temperatures rarely occur in buildings, but are important in beams for bridges. Methods of calculation for shrinkage and temperature effects will be given in Volume 2.

3.9   Partial-interaction design

It sometimes happens in practice that the thickness of a reinforced concrete slab and the size of a steel beam are determined by factors other than their strength when acting together as a composite beam, and that this strength, when calculated by full-interaction theory, is found to be greater than is required. The most economical design may then be one in which the number of shear connectors provided in a half span, N, is such that the degree of interaction between the slab and the steel beam is just sufficient to provide the required flexural strength, and is less than the number N_f required for a full-interaction design.

No provision is made in CP 117 for partial-interaction design of beams, but a method based on allowable stresses was included in the 1969 Specification of the American Institute of Steel Construction. It is similar to the limit-state design method developed for the revised BS 449, which is now described.

It will be evident from Chapter 2 that when the number of shear connectors is reduced below that required for 80% design (p. 51), the effect of slip on stresses and deflections can no longer be neglected. Accurate calculations therefore include partial-interaction theory, which has been shown to be too complicated for everyday use, so an approximate method is required.

The method is applicable only to simply supported beams not subjected to heavy point loads, for it has not yet been verified for other types of beam. The number of shear connectors may not be less than half the number required for full-interaction design, for a small number of shear connectors would be ineffective. This can be seen from the worked example in Section 2.7, in which the maximum slip in the non-composite member at working load was found to be 8·1 mm. If just one connector were placed near each end of the beam (Fig. 2.15) it would shear off at a slip well below 8·1 mm, and so would not increase the flexural strength of the member at all.

It can be deduced from Fig. 3.11(b) that flexural strength at mid-span, M_u, is related to the number of shear connectors, N, as shown by curve BC in Fig. 3.19, in which M_f is the full-interaction strength of the composite member, and M_p is the moment of resistance of the steel member alone. It is therefore safe to use the line DC for design for the ultimate limit state. It gives

$$\left.\begin{array}{l} M_u = M_p + N(M_f - M_p)/N_f \\ \text{where} \quad N \not< 0{\cdot}5N_f \end{array}\right\} \tag{3.37}$$

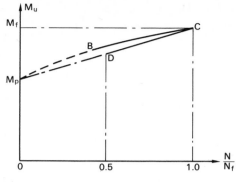

Fig. 3.19

The deflections of beams with incomplete shear connection have been studied by partial-interaction theory.[33] It was found that for a given load, midspan deflection δ is related to the number of shear connectors by a curve similar to EFG in Fig. 3.20, in which δ_f is the deflection given by full-interaction theory and δ_s is that of the steel section alone. It is accurate enough in practice to use the line HJ, which is

$$\delta = \delta_f + \tfrac{1}{2}(\delta_s - \delta_f)[1 - (N/N_f)] \qquad (3.38)$$

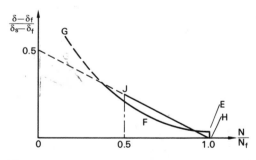

Fig. 3.20

A version of (3.38) more convenient for use in design is

$$\left. \begin{aligned} \delta &= \delta_f + \tfrac{1}{2}\delta_f[(I_c/I_s) - 1][1 - (N/N_f)] \\ \text{where} \quad N &\nless 0{\cdot}5N_f \end{aligned} \right\} \qquad (3.39)$$

and I_c and I_s are the second moments of area of the composite and steel sections, respectively. This result of course relates only to loads carried by the composite member. Deflections of the steel member (e.g., in unpropped construction) are not influenced by the degree of shear connection.

3.9.1 Example (continued). Partial-interaction design

To illustrate the method of Section 3.9, we now study the consequences of halving the number of connectors in the beam used in the design example.

For this beam, $M_f = 770$ kNm (p. 49) and $M_p = 403$ kNm (p. 49). Putting $N = 0.5N_f$ in (3.37),

$$M_u = 403 + 0.5 \times 367 = 586 \text{ kNm}$$

which corresponds to a distributed load of 58 kN/m on a 9 m span. The imposed load corresponding to this reduced flexural strength is now calculated, assuming the dead load to be as before. From p. 44, the factored dead load is 22 kN/m, so the new imposed load is 36 kN/m. This corresponds to a characteristic distributed load on the slab of $36/1.6 \times 4 = 5.6$ kN/m^2, which is 75% of the previous value.

The deflection due to the loading for the serviceability limit state is now estimated. The value of I_c in (3.39) depends on the modular ratio, and for the present beam is 85 000 cm^4 for imposed load and 72 870 cm^4 for dead load. Here it is accurate enough to use the mean value, 78 900 cm^4, for the total load. For the steel member, $I_s = 23\,800$ cm^4, so $I_c/I_s = 3.31$. In Section 3.7.4, the imposed-load deflection of this beam was found to be 15 mm. Scaling this down to 11.3 mm, to allow for the reduced loading, and adding the dead-load deflection for propped construction, 9 mm, gives $\delta_f = 20.3$ mm. From (3.37) with $N = 0.5N_f$,

$$\delta = 20.3(1 + 0.5 \times 2.31 \times 0.5) = 32 \text{ mm}$$

This is 1/280 of the span, and so might be unacceptable in practice unless the steel beams were propped to a camber of about 15 mm.

The accuracy of the design method for this particular beam was checked by analysing the beam at the serviceability limit state by partial-interaction theory. Using results given in Appendix A, and assuming the connector modulus k to be 150 kN/mm, as in Section

2.7, the maximum slip was found to be 0·45 mm. This corresponds to a maximum load per connector of 67 kN, which is sufficiently far below the characteristic push-out strength (100 kN) for the analysis to be valid.

The analysis also showed the curvature at midspan to be 30% greater than that given by full-interaction theory for the same loading. The increase in deflection would be similar. The increase given by (3.39) was 57%, so the approximate method is conservative in this instance.

3.10   Form-reinforced composite slabs

As explained on p. 28, corrugated-metal decking can be used as both formwork and reinforcement for in-situ concrete floor slabs. Two common types of floor are shown in section in Figs. 2.8 and 3.21. There are many variations on these designs, to satisfy various requirements for the incorporation of services within the floor. The ribs in the second type provide convenient anchorage for hangers that carry services below the floor.

Design methods for such floors have been developed by the manufacturers of the decking, and a British code of practice is in preparation. A committee of the European Convention for Constructional Steelwork has been working since 1973 on *European Recommendations for the Calculation and Design of Composite Floors with Profiled Steel Sheet*. These are likely to be based mainly on Swiss[34] and North American[1] work. The available information relevant to the limit-state design of such floors is now summarised and discussed, and a worked example is given.

3.10.1   Properties of cold-formed steel decking

In buildings, form-reinforced slabs are economic only when very thin decking is used. The effects of corrosion on steel sheets about 1 mm thick are more severe than on thicker sections, so the material is usually galvanised. There is evidence that the galvanising process can increase the yield strength of low-carbon steel sheet by as much as 20%. The process of forming the decking from flat sheet causes further localised increases at the corners of the cross-section, due to cold working.

Another important consequence of the use of this sheeting is that the manufacturer's tolerance on thickness will be relatively greater than for hot-rolled steel sections. For example, the value for 1 mm sheet is likely to be ± 0.12 mm, or $\pm 12\%$. This possibility that the material may be only seven-eighths as thick as expected can be allowed for by using a reduced thickness, or more rationally by using a value of the partial safety factor γ_m that takes account of the variability of thickness found in practice. The depth of shallow indentations provided to improve shear connection is subject to a similar variability.[35]

Agreed values of characteristic yield stresses for steel decking and of the partial safety factor γ_m for use in limit-state design of form-reinforced composite slabs are not yet available (April 1975), so that design is normally based on manufacturers' recommendations.

As with standard rolled-steel sections, designers of structures have to select from a limited range of available decking profiles, rather than design their own. The necessary depth of the ribs is determined by the loading when the slab is cast. The decking then has to carry both the wet concrete slab and the imposed loads associated with its construction. It is often found that two or more temporary supports are necessary within each span of the decking, both for strength and to limit the deflection due to the wet concrete, so the decking is usually designed for this loading as a continuous beam. The breadth–thickness ratio of the flat top or bottom of a corrugation may be 100 or more. When in compression, such slender panels begin to buckle at quite low stresses, and for economy, account must be taken of their post-buckling strength. The established theories and design methods for thin-walled steel sections[36] are applicable, but it may be found that panel slendernesses lie outside the range for which design stresses are given in Addendum No. 1 to BS 449.[37] More comprehensive data is available,[34] and was used in the example that follows.

When temporary supports are used the form-reinforced slab has eventually to carry both its own weight and the imposed loads, but even so, the amount of steel in the decking is likely to be more than is needed to reinforce the slab. It is therefore usual to design the slab as simply supported for the ultimate limit state, even though both it and the decking may be continuous over the supporting beams or walls.

For decking of the type shown in Fig. 3.21, bond stress may control the design. In some countries, it is assumed that bond becomes unreliable as soon as the concrete cracks. Design for flexure is then based on the uncracked section, with a limiting tensile stress for concrete. Tests have shown[1] this method to be conservative. The current British practice is to use the conventional *cracked-section* theory for beams subjected to static loading only, with flexural strength calculated either by the elastic method of transformed sections, or by plastic theory. Due to the brittle nature of bond failure, local bond stress due to longitudinal shear should be calculated on an elastic basis.

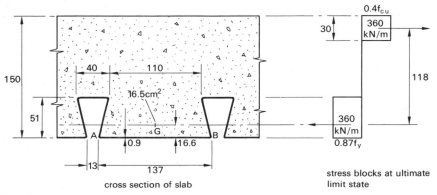

cross section of slab   stress blocks at ultimate
                                             limit state

Fig. 3.21

The choice of a design bond stress for a given decking should be based on a conservative interpretation of the results of tests to failure on form-reinforced slabs made with that decking. Research by Resevsky[38] on the profile shown in Fig. 3.21 gave the following information.

(1) The rigidity of the rib affects bond, so that thicker sheets perform better.
(2) Since there is no keying action to the flat pan (*AB* in Fig. 3.21) bond to it should not be assumed to be fully reliable for long service. (Account is usually taken of this by using a reduced value of mean bond stress, rather than by neglecting bond to a certain area.)
(3) Galvanising slightly improves bond strength.

For profiled sheets without indentations, an allowable (i.e. working-load) bond stress of 0·05 N/mm² (7 lb/in²) has been proposed, calculated on the whole area of contact, unless tests have shown that a higher value can be used.[39] It seems to be generally agreed that for internal spans of a continuous slab, where local bond failure cannot cause collapse, values exceeding 0.05 N/mm^2 are acceptable; but no limiting value for this situation has been established.

3.10.2   Design of form-reinforced slab

Design for flexure is based on the methods used for one-way reinforced concrete slabs, and is straightforward, but a little tedious when elastic theory is used. The main problem is to ensure that there is sufficient shear connection, particularly when the surface of the decking is not deformed or indented. Progressive bond failure could then lead to sudden collapse of an end span, so some positive form of anchorage should be provided, as shown for example in Fig. 3.22(a), in which

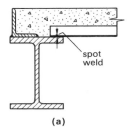
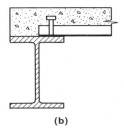

(a)   (b)

Fig. 3.22

tensile force in the decking is transferred by spot welds or other fasteners through the steel section to an angle that resists the corresponding compression in the concrete slab. Shear connectors (Fig. 3.22(b)) provide an alternative method. It is difficult to determine the force for which the anchorage should be designed. If, as intended, it prevents breakdown of bond, then the longitudinal force in the decking at the end of a span should be little more than that due to shrinkage of the slab and any temperature difference between the slab and the decking. But if bond breaks down completely, then before the span fails in flexure, the force applied to end anchorages could reach a value corresponding to yield in tension of the whole cross-section of the metal decking. A calculation based on this assumption is given on p. 96. More research is needed on this subject, and also on the amount of anchorage required near internal supports and on the extent to which repeated or impulsive loading in buildings accelerates the breakdown of bond.

In a series of static and fatigue tests[35] on composite slabs of the type shown in Fig. 2.8, data relevant to the design of form-reinforced bridge decks was obtained. The work showed that longitudinal shear strength is much improved by indentations or ribs that create positive interlock between the steel form and the concrete slab. It was concluded that when designing for fatigue loading, no reliance should be placed on bond, and that resistance to longitudinal shear failure can be assumed to be proportional to the amount of interlock provided. This interlock is obviously created by the dimples shown in Fig. 2.8, but the effect of the re-entrant shape of the corrugations in Fig. 3.21 is less well understood. The worked example will show that the strength of a slab with this type of decking can be limited by bond stress, even when no account is taken of repeated loading. It is likely that even a slight variation in the cross-section of the decking in the longitudinal direction, for example by crimping or indenting its surfaces, would significantly improve its behaviour, and so eliminate any need for end anchorages.

The serviceability limit state that most influences design and the method of construction is excessive deflection. The total deflection is the sum of three components:

(1) deflection of the decking between temporary supports, due to the weight of the concrete slab, plus any deflection of the supports themselves;
(2) elastic and creep deflection of the composite slab due to the load initially carried by the temporary supports, and due to finishes and other permanent loads;
(3) elastic deflection of the composite slab due to imposed load.

In theory, account should be taken in calculations for metal decking of the reduction in the effective width of a slender plate panel due to the buckling mentioned above. Accurate calculation of deflection would then be a long process. In practice, this effect can usually be neglected, particularly when no account is taken of the beneficial effect of continuity of the slab over the supports.

It is usual to provide square-mesh steel reinforcement at or above mid-depth of the slab to minimise cracking due to shrinkage and temperature effects, and to help to distribute concentrated loads. This reinforcement should be in accordance with the requirements of CP 110[9] for secondary reinforcement in slabs. It serves also to control the width of cracks where the slab is continuous over supports. Increasing the amount of reinforcement over supports is an effective method of reducing midspan deflections.

In detailing reinforcement for form-reinforced slabs, it should be remembered that the ability of the slab to distribute the effects of local point loads may be less than that of a reinforced concrete slab of the same overall thickness, particularly near a hole or an edge of a form-work panel.

The following calculations show how design methods from CP 110, CP 117: Part 1[11] and the Swiss Recommendations[34] can be applied to form-reinforced slabs, and illustrate the preceding discussion.

3.10.3   Example

In the example of Section 3.3.1, the floor slab was of in-situ concrete, 150 mm thick, and was continuous over a series of 4-m spans. This slab is now re-designed using permanent steel formwork of the type shown in Fig. 3.21. It is assumed that the overall thickness and the design loads (p. 44) are as before.

In practice, the design of such a slab might consist simply of looking up data tabulated in a manufacturer's catalogue. But the designer should understand how such data is obtained, and for an unusual problem may need to do calculations similar to those given here.

The unfactored imposed loading including finishes is therefore 7·5 kN/m², and the slab thickness is 150 mm. The suitability of the cross-section shown in Fig. 3.21 will be examined. The galvanised steel decking is *Holorib*, available from Richard Lees Ltd in widths of 0·6 m and lengths sufficient to extend over two 4-m spans. As explained on p. 81, limit-state design parameters for this material were not available in April 1974, so it will be assumed that the characteristic mean yield stress of the decking is 250 N/mm², with $\gamma_m = 1\cdot15$ at the ultimate limit state.

The properties of a 1-m width of the decking, in the usual notation, are given by the manufacturers as follows:

$$A_s = 16.5 \text{ cm}^2 \qquad I = 67 \text{ cm}^4 \qquad Z = 40.3 \text{ cm}^2$$

for surface AB (Fig. 3.21). The centroid is 16·6 mm above the surface AB, and the area of the steel–concrete interface is 1·91 m²/m.

Stresses and deflections during construction. The calculations are made for the serviceability (unfactored) loads, because these are the loads for which deflections are required. Based on the Report *False-work* (Concrete Society, London, July 1971), the imposed load during concreting is taken as 2 kN/m². Adding the weight of the slab, 3·6 kN/m², gives a total load, w, of 5·6 kN/m².

The span–depth ratio of the decking alone is 4/0·051, or 79, so it is obvious that temporary supports are needed. Use two at the one-third points reduces the span–depth ratio to 26, which should be low enough. To allow for some inaccuracy in locating these supports, we assume that the spans between them are 1·4 m, rather than 1·33 m (Fig. 3.23). The maximum bending moments in a continuous strip of slab 1 m wide are given by Clause 3.3.4 of CP 110 as $+wL^2/11$ and $-wL^2/9$, or $+1.0$ kNm and -1.22 kNm, as shown in Fig. 3.23. The bottom-fibre tensile stress at point B is

$$f_{bt} = 1000/40.3 = +24.8 \text{ N/mm}^2 \qquad (3.40)$$

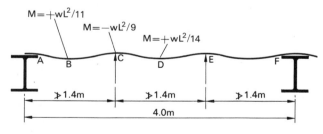

Fig. 3.23

Over a support, plate AB (Fig. 3.21) is in compression, and allowance must be made for its slenderness. This is $136/0.9 = 150$, which is outside the range of data given in PD 4064,[37] so the Swiss Recommendations[34] are used instead. For a compression flange supported by a web at each edge, and of slenderness b/t, these give the relationship between the effective breadth b_e of the compressed plate and the working-load stress at its edges, f_c (in N/mm^2 units), as

$$\frac{b_e}{t} = \frac{856}{(f_c)^{1/2}}\left[1 - \frac{188t}{b(f_c)^{1/2}}\right] \tag{3.41}$$

The value of f_c for a given bending moment depends on the section modulus of the decking, which itself depends on the effective breadth b_e, so a tedious trial-and-error calculation is required for f_c. The solution in this case is

$$f_c = -40 \text{ N/mm}^2 \qquad b_e/t = 109 \tag{3.42}$$

For a slender plate, the relation between bending moment and stress f_c is non-linear, due to the variation of effective breadth with stress; but the value of f_c found above is so far below the yield stress (250 N/mm^2) that it is obvious without calculation that yield would not occur under the factored construction loads. An explanation of the use of effective breadth in the analysis of thin compressed plates is given in Reference 36.

The maximum total deflection will occur near point D (Fig. 3.23). If the supports at C and E do not settle, the span between them can be treated as fixed-ended, so its deflection due to the concrete slab is $wL^4/384EI$, where $w = 3\cdot6$ kN/m^2, $L = 1\cdot4$ m, $E = 200$ kN/mm^2, and $I = 67$ cm^4. This gives a deflection of only $0\cdot3$ mm, which can be neglected.

Deflection in service. In calculating the other components of the total deflection at D, the span AF is assumed to be simply supported. A simple and conservative way of allowing for the removal of the props at C and E is to assume that the whole of the self-weight is carried by the composite slab. To allow for creep, the modular ratio is taken as 15, and the elastic properties of a 1-m width of slab (Fig. 3.24) are calculated in 'steel' units by the method of transformed sections, exactly as for a composite beam.

The neutral-axis depth x is found to be 60 mm, and the second moment of area of the cracked section, I, is 1430 cm^4. The deflection of the slab due to its own weight of $3\cdot6$ kN/m^2 is $5wL^4/384EI$:

$$(5 \times 3\cdot6 \times 4^4 \times 10^5)/(384 \times 200 \times 1430) = 4\cdot2 \text{ mm} \tag{3.43}$$

It is simple and conservative also to use $m = 15$ for the imposed load of 7.5 kN/m^2. From (3.43), this causes a deflection of $4.2 \times 7.5/3.6 = 8.8$ mm, giving a total of 13 mm, or span/308, which is below the recommended limit[9, 15] of span/250. If this limit were found to be exceeded, more precise calculations would be worth while. For example, the imposed-load deflection is reduced from 8.8 mm to 7.2 mm if m is taken as 7.5, and the crack-control reinforcement over the supports would cause a further reduction.

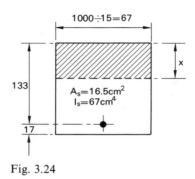

Fig. 3.24

For propped construction, a simpler and quite accurate method of checking for excessive deflection would be to use the limiting span–depth ratios given in CP 110, taking the effective depth of the slab as the depth to the centroid of the steel decking.

Ultimate limit state.   Local bond stress will be checked first. The limiting value recommended by the manufacturers of Holorib (0.07 N/mm^2) relates to working loads and calculations on an elastic basis. It is on the safe side here to use $m = 15$ for all loads. The total vertical shear at one end of a simply-supported span is

$$V = 2(3.6 + 7.5) = 22.2 \text{ kN/m}$$

Of this, the decking alone resists the shear due to dead load on a span of about 1.4 m to the nearest prop. This shear is $0.7 \times 3.6 = 2.5 \text{ kN/m}$, so the design shear for the composite slab is $22.2 - 2.5 = 19.7 \text{ kN/m}$. The shear stress at the interface, v, is given by

$$v = V A_s \bar{y}/Ib$$

where A_s is the area per unit width of the steel decking, $\bar{y}$ is the depth of its centroid below the elastic neutral axis ($133 - 60 = 73$ mm) and b is the length of the steel–concrete interface ($1\cdot91$ m^2/m, from p. 86). Thus,

$$v = (19\cdot7 \times 16\cdot5 \times 73)/(1430 \times 1\cdot91 \times 100) = 0\cdot087 \text{ N/mm}^2$$

This exceeds the limiting value of $0\cdot07$ N/mm^2. It might be acceptable in internal spans, but it would be necessary to provide anchorage (Fig. 3.22) or some additional form of shear connection in an end span. An example of this is given at the end of this section.

The flexural strength of the slab is now checked. This rarely governs, so it is worth making approximations, particularly when elastic theory is used. First, the stresses due to the working loads are compared with the limiting stresses given in CP 117: Part 1. It is sufficiently accurate to use $m = 15$, and to assume that all loads are carried by the composite section, with a uniform tensile stress in the metal decking. In the present example, this stress is 119 N/mm^2 at midspan. To allow for the stresses due to the method of construction it is on the safe side to add to this stress the tensile stress in the decking due to the moment $wL^2/14$ at midspan (Fig. 3.23), where w is the weight of the wet concrete ($3\cdot6$ kN/m^2). This stress is

$$3600 \times 1\cdot4^2/14 \times 40\cdot3 = 13 \text{ N/mm}^2$$

giving a total of 132 N/mm^2, or $0\cdot53f_y$. This is well below the limiting stress of $0\cdot9f_y$ given in CP 117, so that these approximate calculations are accurate enough. The maximum compressive stress in the concrete slab is $9\cdot8$ N/mm^2, which is also satisfactory.

Alternatively, the flexural strength can be checked at the ultimate limit state. If the plastic neutral axis is above the top of the decking, as is usual, the calculation is the same as for a reinforced concrete slab. The design charts in CP 110: Part 2 can be used, provided that allowance is made for any difference between the partial safety factor γ_m for steel in metal decking (not known at the time of writing) and that for reinforcing steel. In this method, no account need be taken of the method of construction.

The factored loading is

$$1\cdot4 \times 3\cdot6 + 1\cdot6 \times 7\cdot5 = 17\cdot0 \text{ kN/m}^2$$

The design midspan moment is $17 \times 4^2/8 = 34$ kNm. We assume that $\gamma_m = 1.15$ for steel, $f_y = 250$ N/mm^2, and $f_{cu} = 30$ N/mm^2, and that the depth of the concrete stress block may not exceed half the depth to the steel centroid, or 67 mm. The available force in the steel is $0.87 \times 250 \times 1.65 = 360$ kN/m, (Fig. 3.21), and the corresponding depth of the stress block in the concrete is $360/(0.4 \times 30) = 30$ mm. The lever arm is $133 - \frac{1}{2} \times 30 = 118$ mm, so that

$$M_u = 360 \times 0.118 = 42.5 \text{ kNm}$$

which well exceeds the 34 kNm required.

These calculations show that the proposed design is adequate in all respects except bond strength in an end span, where anchorage should be provided. This is now designed by the method given on p. 83. An estimate of the force to be resisted can be made by scaling down the available force in the steel (360 kN/m) by the ratio of the required strength to the calculated strength (34/42.5). This gives 288 kN/m at the ultimate limit state. If the slab is also acting as the flange of the steel beam supporting the decking (Fig. 3.22(b)) it may be convenient to provide this anchorage by means of shear connectors, since these will be effective in both the longitudinal and the transverse directions. Their size and spacing is therefore calculated on p. 96, after the design of such beams has been discussed.

3.11   Composite beams supporting form-reinforced slabs

A form-reinforced slab can conveniently be attached to a supporting steel beam by means of shear connectors welded through the decking, which may be laid continuously over the beam. This process is now standard practice in North America, where it has been found that reliable welds can be made provided that the thickness of the decking and its surface finish lie within established limits.[1] The slab can thus act as the top flange of a composite beam; but the corrugated shape of its underside influences the design of the beam. The two arrangements that arise in practice are now considered in turn.

3.11.1   Slab with ribs parallel to the steel beam

When the slab spans in the same direction as the beam (Fig. 3.25), it is not essential to weld the connectors through the decking; but if this is not done the decking should be attached to the steel beam, for example by stud welding, sufficiently strongly to enable it to act as bottom transverse reinforcement for the beam, and so to prevent shear failure on planes *ABCD* or *EBCF*.

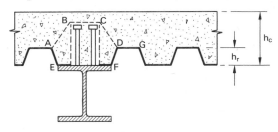

Fig. 3.25

In designing the beam, the effective breadth of the slab can be assumed[1] to be the same as for a solid slab of depth h_c; but if the depth of the concrete stress block, x, exceeds the thickness of concrete above the decking $(h_c - h_r)$, allowance must be made for the missing concrete between the ribs of the decking. It is rarely worthwhile to allow for the strength of the decking when designing the beam, for several reasons: the decking is close to the neutral axis of the composite member; except near the beam, it is already stressed by the weight of the concrete slab; and account must be taken of the effect of buckling on the compressive strength of a slender plate such as *DG* in Fig. 3.25.

Consideration should be given to the combination of stress in the concrete due to slab bending and to beam bending, as both span in the same direction; but in this and other respects the design method is as for a conventional composite T-beam.

3.11.2   Slab with ribs at right angles to the steel beam

With this arrangement (Fig. 3.26), the connector spacing is determined by the rib spacing, and there are voids above the steel beam and between the connectors. Current recommendations[34] for design are applicable only when the dimensions of the slab and connectors lie within the limits shown in Fig. 3.26, which are based on the available research.[1] These references include information on design with lightweight-aggregate concrete. For brevity, only normal-density concrete is considered here.

It should be assumed that the effective breadth of the flange of the beam is as for a solid slab. Use of a lower value may lead to over-loading of the connectors.

Rigorous calculations based on elastic theory are often tedious, for the elastic neutral axis of the composite beam usually lies within the depth of the decking. For this reason the following simplifications are given in terms of y, the depth of the compressive stress block given by simple plastic theory, neglecting the concrete below the top of the decking.

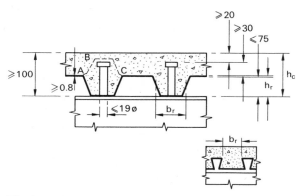

Fig. 3.26

If $y \leqslant (h_c - h_r)$ *and* $h_r \leqslant 40$ mm, the effect of the voids on *elastic* behaviour can be neglected, and deflections and working-load stresses calculated as if the slab were solid.

Otherwise, the concrete below the top of the ribs should be neglected when calculating stresses in concrete and deflections. The missing concrete has little effect on the stress in the bottom flange of the steel beam. Approximate expressions are available[1] for the moment of inertia and top-flange section modulus of the beam.

The most important effect of the voids is to reduce the strength and stiffness of the shear connection. Tests show that the strength of the beam depends on the geometry of the ribs, as well as on the number of shear connectors, because cracks occur at the corners A and C (Fig. 3.26), and may lead to shear failure on surface ABC. In narrow slabs of the type used in push-out tests, this failure may extend to the edges of the slab, and then the strength of a rib is found to increase with slab breadth. But in T-beams, where the slab is effectively very wide, the failure is local and independent of slab breadth. It places a limit on the shear strength of a rib, that is independent of the number of connectors used.

It was deduced from push-out tests[1] that P_{ur}, the strength of a shear connector in a rib of breadth b_r and height h_r, is given by

$$P_{ur} = 0.36(b_r/h_r)P_u \qquad (3.44)$$

where P_u is its strength in a solid concrete slab. In re-entrant ribs (inset on Fig. 3.26), b_r should be taken as the minimum breadth; otherwise the mean breadth can be used. In beams at flexural failure, the loads on connectors are on average less than is given by full-interaction theory. Fisher's study of tests on beams showed that the design strength was always reached when b_r/h_r exceeded 2·0, so that design could be based on

$$P_{dr} = 0.5(b_r/h_r)P_d \qquad \text{(but } P_{dr} \not> P_d) \qquad (3.45)$$

where P_{dr} and P_d are the design strengths of a connector in a rib and in a solid slab, respectively.

Fisher concluded that (3.45) is valid for ribs in which there are not more than two $\frac{3}{4}$-inch (19-mm) stud connectors, or the equivalent area of smaller sizes. Of the test beams in which full-interaction design was used, none had a rib breadth greater than 127 mm, so further testing may show that a higher limit is appropriate for wider ribs. In the absence of data from tests on L-beams, it would be prudent to limit the shear connection provided in such beams to one 19-mm stud per rib, or its equivalent.

When the decking has ribs with a low b_r/h_r ratio, it may not be possible to fit in sufficient shear connection for the beam to be designed for full interaction; and there may be other situations in which partial-interaction design (p. 76) reduces cost. For simply supported beams at the ultimate limit state, the following procedure is suggested for checking the strength and designing the shear connection for a beam of given cross-section.

(1) Neglecting concrete below the top of the ribs, calculate the depth of the plastic neutral axis for full interaction, and the corresponding plastic moment of resistance, M_u.

(2) Calculate the number of shear connectors required for full interaction, N_f, and provide these, if possible, unless the flexural
strength required is much less than M_u. In this event, or if the full
number of connectors cannot be fitted in, use partial-interaction
design. Then Equation (3.37) (p. 77) can be used to find either
the number of connectors needed for a given flexural strength, or
the flexural strength when the maximum possible number of
connectors is used.

This method is illustrated in Section 3.11.3.

When partial-interaction design is used, it may not be necessary to
place a connector in every rib. Stud spacings up to 660 mm have been
used in test specimens.[1] It was found that when there are ribs without
connectors, the stiffness of the beam is slightly reduced, and that the
reduction increases with height of rib. It is therefore recommended
that unless shown to be satisfactory by tests on beams, ribs without
shear connectors should be spot-welded or otherwise attached to the
steel beam.

3.11.3   Example (continued)

The 9-m span simply supported beam studied in Section 3.4.2 (p. 48)
is now redesigned to support the form-reinforced slab designed in
Section 3.10.1.

From p. 44, the required moment of resistance at the ultimate
limit state is 709 kNm, and the strength of the beam with a solid slab
is 770 kNm. We therefore try the same steel section (Fig. 3.27) even

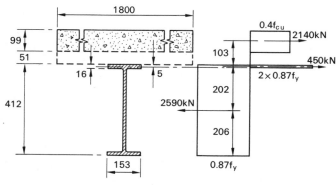

Fig. 3.27

though the available depth of concrete in compression is reduced from 150 mm to 99 mm (Fig. 3.21). The effective breadth is 1·8 m as before, so the available compressive force in the concrete is

$$0·4 \times 30 \times 1·8 \times 99 = 2140 \text{ kN}$$

From p. 48, the steel I-section is capable of resisting a force of 2590 kN, so the plastic neutral axis is likely to lie within the top flange, at a depth z (say) given by

$$2 \times 0·87 \times 350 \times 0·153z = 2590 - 2140$$

whence

$$z = 4·8 \text{ mm, say 5 mm}$$

and

$$M_u = 2140 \times 0·103 + 2590 \times 0·202 = 743 \text{ kNm}$$

This exceeds 709 kNm, so the beam will be strong enough if the shear connection can be provided.

From Fig. 3.21, the ratio b_r/h_r for a rib is 110/51, or 2·16. This exceeds 2·0, so from (3.45) the full design strength of the connectors can be used. Studs 19 mm in diameter and 100 mm high lie within the limits given in Fig. 3.26, and will be used. Their design strength (p. 54) is 69·5 kN per stud, so the number needed in a half span is 2140/69·5, or 31.

The ribs are spaced 0·15 m apart, so the number in a half span is 4·5/0·15, or 30. It is therefore convenient to design for a small degree of partial interaction and provide one stud in each rib. The reduction in the strength of the beam due to the use of 30 studs rather than 31 is found from (3.37), with $M_p = 403$ kNm (from p. 49):

$$M_u = 403 + 30(745 - 403)/31 = 732 \text{ kNm}$$

This also exceeds the required strength of 709 kNm.

In the Swiss Recommendations,[34] it is suggested that when the plastic neutral axis lies below the top of the ribs, as here, the moment of resistance of the beam can be assumed to be 15% less than that if the slab were solid. The reduction here is from 770 to 732 kNm (5%), and the rule, if applied here, would have led to the conclusion that the beam was not strong enough.

The deflection of this beam should be calculated by conventional elastic analysis of the cross-section shown in Fig. 3.27. The increase due to the use of partial-interaction design (p. 78) would be negligible in this example.

To illustrate the use of studs both as shear connectors in the longitudinal direction and as end anchorages in the transverse direction, we now consider an end span of the form-reinforced slab (Fig. 3.22(b)) and assume that the loading on the composite edge beam is such that the compressive force in the concrete flange at midspan is half the value for an internal beam, or 1070 kN, which is 238 kN/m over the half span of 4·5 m. It was found on p. 90 that the required anchorage force for the form-reinforced slab is 288 kN/m. The resultant force is therefore $(238^2 + 288^2)^{1/2}$, or 373 kN/m. Using 19-mm studs as before, their spacing should not exceed 69·5/373 m, or 0·19 m. In practice it would be convenient to provide one per rib (pitch 0·15 m).

3.12   Vibration of long-span composite floors

The use of composite action makes possible the construction of floors that are very light in relation to their span, particularly if metal decking is used as permanent formwork. Some long-span floors supporting open areas without partitions have been found to vibrate under normal usage to an extent that causes complaints. It is therefore prudent to check the 'liveliness' of such floors at the design stage. A method of doing this has been developed in North America.[40] In it, the structure is assumed to consist of parallel simply supported composite beams, and the check is made for an individual beam, neglecting interaction between it and those on either side. Steady vibration, due for example to heavy machinery, is not considered, for this should be isolated at source, and the structure designed to avoid resonance effects.

The present problem is transient vibration due to persons moving about in normal occupancy. Neither the types of impact causing vibration nor the human reactions to it can be precisely quantified, so any check of this kind is bound to be approximate. The initial disturbance is assumed to be a 'heel drop', caused by a man weighing about 0·75 kN standing on the balls of his feet with his heels about

65 mm above the floor, and then relaxing, so that his heels impact the floor. The frequency of the transient vibration so caused is assumed to be the fundamental natural frequency of the floor, f_1. This is calculated from the standard expression derived by elastic theory for a simply supported uniform undamped beam of span L,

$$f_1 = \frac{\pi}{2}\left(\frac{gE_sI}{WL^3}\right)^{1/2} \tag{3.46}$$

where I is the second moment of area of the uncracked composite section, calculated using the full flange breadth rather than the effective breadth, and W is the total weight of the beam, including finishes and partitions.

It has been found that human response to a transient vibration is strongly influenced by the rate at which it dies away. If it continues for twelve cycles or more, the response is as for a steady vibration. If there is sufficient damping for the amplitude after five cycles to have fallen to less than 20% of the initial amplitude, then for a given floor structure, the initial amplitude can be ten times as great as that of a steady vibration, for the same degree of human perception or annoyance. This conclusion has been used with data on strengths of sensations of standing persons subjected to uniform vertical vibration to obtain Fig. 3.28. This gives the degree of damping needed in floors of various natural frequencies (f_1) for transient vibrations of various initial amplitudes (A) to reach the *threshold of annoyance*. This threshold corresponds roughly to the boundary between the human responses *not perceptible* and *slightly perceptible*.

There is no merit in accurate calculation of the initial amplitude when it is due to a disturbance as approximate and arbitrary as the 'heel drop'. Allen reports[40] that the following empirical expression gives good agreement with test data from about thirty floors, and is applicable to beams with natural frequencies less than about 10 Hz (i.e., spans exceeding about 7·5 m):

$$A = \frac{0\cdot17}{f_1 L h_c (h_c + 0\cdot025)} \tag{3.47}$$

where A is the initial amplitude in millimetres, f_1 is the natural frequency in Hz, L is the span in metres, h_c is the thickness of the concrete slab (including ribs if any) in metres. When lightweight-aggregate concrete is used for the slab, the constant 0·17 should be replaced by 0·22.

At the design stage, the degree of damping inherent in a floor can only be estimated. It is expressed as the ratio of the damping to the critical damping for the structure considered. Figure 3.28 shows that vibration is unlikely to be troublesome if the damping ratio exceeds 0·10. The following data, based on a study of thirty floors[40] should be applicable to one-way composite floor systems generally.

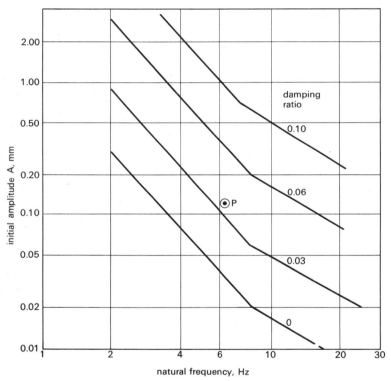

Fig. 3.28 (Courtesy the Canadian Steel Industries Construction Council)

The four curves relating damping ratio to span shown in Fig. 3.29 apply to the following types of floor construction:

(A) 65 mm (minimum) concrete slab on 38-mm metal decking, with I-section or open-web steel beams. No suspended ceiling, partitions, floor finish, air-conditioning ducts, or electrical conduits.

(B) As (A), but with suspended ceiling, ducts and conduits; floor finished and carpeted, but no partitions.
(C) As (B), but with 120-mm concrete floor slab on steel beams.
(D) All types of floor construction with ceiling-height partitions, not necessarily connected to the floor above.

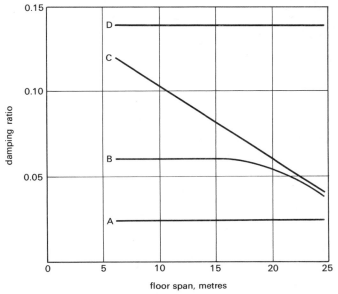

Fig. 3.29 (Courtesy the Canadian Steel Industries Construction Council)

It has also been found that:

(1) Increase of slab thickness from 65 mm to 90 mm raises curves (A) and (B) by about 0·01 for spans up to 15 m, and less at longer spans.
(2) Partial-height lightweight partitions appear to give a damping ratio of at least 0·10, for all spans.
(3) The 'heel drop' is an inadequate measure of the force inputs that can occur from activities in dance halls and gymnasia. There is little data on acceptability criteria for such floors. Based on tests in two dance halls, Allen suggests that the calculated value of (A) should be multiplied by 1·5 to 2·0 before using Fig. 3.28. These floors are usually finished but unfurnished, with a suspended ceiling and minimal ducts and conduits. Floors of type (A) construction can be assumed to have a damping ratio of about 0·035.

It is no easy matter to reduce the liveliness of a floor once it has been constructed. Allen gives examples where this has been done by increasing the damping ratio.

3.12.1   Example. Vibrational behaviour of the floor structure

Of the three alternative floor structures designed in this chapter, the one using metal decking (Section 3.10) will have the lowest damping ratio. Its response to transient loadings is therefore examined.

In calculating the second moment of area of a typical beam, we assume that the steel decking compensates for the missing concrete between ribs, and use a flange breadth of 4 m (the beam spacing) and the short-term modular ratio (7·5). For the uncracked section, it is found that $I = 99\,600$ cm^4.

Vibration is a serviceability problem, so γ_f for loads is taken as 1·0. From p. 70, the dead load is 15 kN/m. We assume that finishes (mainly a concrete screed) account for a further 5 kN/m (included with imposed load in earlier calculations), so the total load on a beam is $20 \times 9 = 180$ kN. From (3.46),

$$f_1 = 1\cdot57(9\cdot81 \times 2 \times 99\,600/180 \times 9^3)^{1/2} = 6\cdot1 \text{ Hz}$$

From (3.47),

$$A = 0\cdot17/(6\cdot1 \times 9 \times 0\cdot15 \times 0\cdot175) = 0\cdot12 \text{ mm}$$

These two values give point P on Fig. 3.28, which shows that the threshold of annoyance is reached if the damping ratio is about 0·04.

The effective thickness of concrete in this floor is little less than 150 mm, as the voids between ribs are small. From (1) on p. 99, 25 mm of extra concrete increases the damping ratio by 0·01. This floor is about 50 mm thicker than floor (A), and so should have a ratio of at least 0·045 even if there are no partitions, false ceiling, ducts, or floor finish. With a 50-mm floor finish (as assumed), the damping is likely to be between curves (B) and (C) in Fig. 3.29, and so well above the 0·04 required.

CHAPTER FOUR

Continuous Composite Beams

4.1 Design methods for beams in buildings

The only design method for continuous composite beams given in current British codes of practice is that of CP 117: Part 2.[12] It requires longitudinal moments and the corresponding flexural and shear stresses to be calculated by the elastic theory, assuming full interaction and that concrete has no strength in tension. These stresses must not exceed the relevant working stresses for steel and concrete given in BS 153[41] and CP 114.[42] The calculations required are quite extensive, for these reasons:

(1) Allowance for creep of concrete involves the use of two modular ratios, one for permanent (dead) loads and one for transient (imposed) loads. Two different values of the second moment of area of the uncracked composite cross-section are therefore required.

(2) The longitudinal tensile stress in the concrete slab in negative-moment regions is influenced by the sequence of construction of the deck slab and the method of propping used, and by the effects of temperature and shrinkage, as well as by the dead and imposed loading. Thus the loss of stiffness in these regions due to cracking of concrete cannot accurately be predicted. The flexural

rigidity of a 'cracked' composite cross-section can be as low as a quarter of the 'uncracked' value, so that a wide variation in flexural rigidity can occur along a composite beam of uniform cross-section, leading to uncertainty in the distribution of longitudinal moments and hence in the amount of cracking to be expected. This 'closed-loop' problem is solved in CP 117: Part 2 by assuming in calculations that the concrete is either uncracked or fully cracked, depending on which leads to the higher value of the stress being calculated. The design procedure can involve the analysis of continuous beams of non-uniform cross-section, and the calculation of second moments of area for cracked as well as for uncracked sections.

(3) Elastic analysis of moments in one span involves the consideration of loadings on adjacent spans and the construction of envelopes of maximum positive and negative moments for the span. When unpropped construction is used, there are three such envelopes: those due to dead load carried by the steel section and to dead and to imposed load carried by the composite section. All three have to be kept separate because the relevant section properties are different.

Experience has led to some simplification of these calculations, and the design method of CP 117: Part 2 has been found to be acceptable in practice for bridge beams. But the cost of a continuous beam in a building may be a fifth or a tenth of that of a bridge beam, and there is correspondingly less time available for its design. The need for simpler methods has increased with the change to limit-state design, which involves separate checks at the ultimate load and at working load.

These problems have led to the development of an ultimate-strength design method. It is based on simple plastic theory, as used for rigid-jointed steel structures, and will be described in the following pages, with frequent reference to a worked example. It is hoped that the new BS 449 will allow the use of a method of this type for continuous composite beams. Its main advantages over the method of CP 117: Part 2 are that plastic analysis of both members and sections is much simpler than elastic analysis, that dead and imposed load can be treated alike, and that there need be no consideration at the ultimate limit state of the effects of creep of concrete or of the method of construction.

It will be shown that the plastic method is not applicable to beams in which the slenderness of the web or bottom flange of the steel section exceeds certain limits. Bending moments and stresses in these beams at the ultimate limit state must be calculated by elastic theory. This will automatically ensure that stresses at the serviceability limit state are not excessive.

In beams designed by the plastic method, the elastic theory should be used to check that at the serviceability loads, the stresses in structural steel and reinforcement are not excessive. This subject is considered in Section 4.5.3.

4.2   The design example

So that use can be made of previous work, the design problem is identical with that of Chapter 3, except that the building (Fig. 3.1) now consists of two bays each of span 10·5 m. The transverse beams at 4 m centres are assumed to be continuous over a central longitudinal wall, and are simply supported on columns in the outer wall. Thus each beam is as shown in Fig. 4.1. The use of continuity should offset the increase in span from 9 m to 10·5 m, so it is assumed initially that the design of the slab and the midspan region of the beam are as before (Figs. 3.7 and 3.15), with the same materials, loads, and partial safety

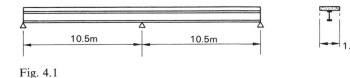

Fig. 4.1

Table 4.1. Loads and bending moments for a span of 10·5 m

	Serviceability limit state		Ultimate limit state	
	Load	$wL^2/8$	Load	$wL^2/8$
Dead load, g	15	207	22	303
Imposed load, q	30	414	48	662
Total load, w	45 kN/m	621 kNm	70 kN/m	965 kNm

factors (p. 41). The design loads per unit length of beam and the corresponding values of $wL^2/8$ for a span of 10·5 m are therefore as given in Table 4.1.

4.3   Negative-moment regions of continuous composite beams

A simple example of a negative-moment region occurs in an isolated T-beam with an in-situ slab that is continuous over two spans. Many such beams have been tested to failure, not only with loads applied over the beam web, but also with line loads along the edges of the slab[24] to simulate the negative transverse bending that occurs when the slab is continuous in the transverse direction over several similar beams, as in a beam-and-slab bridge deck.

When rigid beam–column joints are used, the presence of a steel or composite column (Fig. 4.2) makes little difference to the behaviour of the beam except when the members form part of a frame that resists lateral forces (discussed in Section 5.6). Semi-rigid joints are considered in Section 5.5.

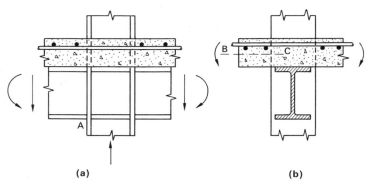

(a)   (b)

Fig. 4.2

When the beams are widely spaced in relation to their span, a two-way composite floor system may be used, with each internal column supporting two continuous beams that intersect at right angles; and a similar situation arises in bridge decks where two widely spaced box girders support closely spaced cross girders, and both are composite with the deck slab. Near a column or pier, the slab is then subjected to biaxial tension. Little research has been done on such slabs, for which empirical design methods are used. The following discussion relates mainly to one-way systems, in which the neutral axis for transverse bending of the slab (*BC* in Fig. 4.2(b)) lies within the slab.

The beams are assumed to be uncased, for little is known of the ultimate strength of negative-moment regions where the steel section is encased in concrete. In particular, no design method is yet available for deciding how much reinforcement is required in the casing at point *A* (Fig. 4.2(a)) to ensure that the concrete in this region, where the compressive strain is very high, can continue to resist compression until the design ultimate load of the beam is reached.

4.3.1 Negative moment of resistance

If the slab is continuous past a support, it is essential to provide longitudinal reinforcement to control cracking of the concrete. In non-composite construction with simply supported steel beams, the light mesh normally used is likely to yield before working load is reached, and its strength is neglected in design. In continuous composite beams, it is economic to provide heavier bars, and to take account of the contribution they make to the negative moment of resistance.

Research on shear lag in negative-moment regions has shown[24] that when transverse reinforcement appropriate to the shear-connector spacing is provided (Section 4.3.7) the slab is well able to transfer shear to longitudinal bars placed six to eight slab thicknesses either side of the steel beam. This in-plane shear causes the diagonal cracks shown in Plate 5. Where heavy longitudinal reinforcement is used, it should be placed fairly close to the steel beam, and account must be taken in the detailing of any holes through the slab in its vicinity. In continuous beams, the effective breadth of the flange may be assumed to be as given by CP 110 for reinforced concrete T-beams (one-fifth of the length of the positive-moment region). All properly anchored longitudinal reinforcement within this breadth may be assumed to be effective, although it is not usually worth including the small bars used to control the spacing of the bottom transverse reinforcement. More detailed rules for effective breadth in steel and composite bridge beams are available.[43]

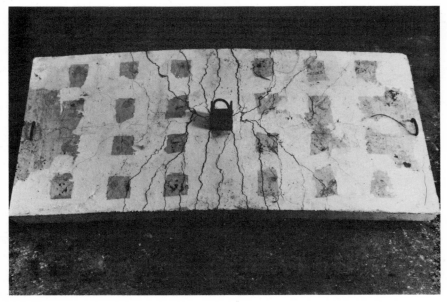

Plate 5. Severe cracking near an internal column in the concrete top flange of a composite beam loaded to failure (courtesy Dr K. Van Dalen)

The following method of calculating the negative moment of resistance M_p' avoids the complexities that arise in some methods when the cross-section of a rolled steel member is assumed to consist of three rectangles. Figure 4.3(a) shows the cross-section considered. The slab reinforcement within the effective breadth, of area

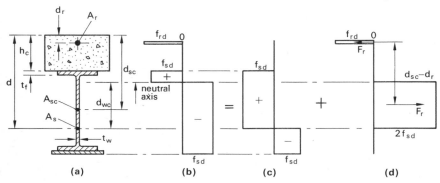

Fig. 4.3

A_r and design yield stress f_{rd} (i.e., f_{ry}/γ_m) is located distance d_r below the top of the slab. Other notation is shown in Fig. 4.3 or is as used in Chapter 3. At flexural failure, the whole of the concrete slab may be assumed to be cracked, and simple plastic theory is applicable, with all the steel at its design yield stress, either in tension or compression. The plastic neutral axis may be in the top flange or in the web, and is first assumed to be in the web, distance d_{wc} above the centroid of the steel section. The stresses are as shown in Fig. 4.3(b), and are split into two sets; those in Fig. 4.3(c) which correspond to the plastic moment of resistance of the steel section alone, M_{ps}, and those in Fig. 4.3(d). Longitudinal equilibrium is satisfied in each of these sets, so the longitudinal forces in Fig. 4.3(d) are as shown, where

$$F_r = A_r f_{rd} \tag{4.1}$$

Then d_{wc} is given by

$$2d_{wc}t_w f_{sd} = A_r f_{rd} \tag{4.2}$$

If $d_{wc} \leqslant d - h_c - t_f$, this result is correct. Then $d_{sc} = d - \frac{1}{2}d_{wc}$ and

$$M_p' = M_{ps} + F_r(d_{sc} - d_r) \tag{4.3}$$

which shows clearly the influence of the reinforcement A_r on the strength of the member.

If $d_{wc} > d - h_c - t_f$, the neutral axis lies in the top flange. Exact calculation is then tedious, and it is on the safe side and often accurate enough to assume

$$d_{sc} = h_c + t_f \tag{4.4}$$

and to use (4.3) to calculate M_p'.

4.3.2   Use of simple plastic theory for composite beams

Accounts of plastic theory and the design methods based on it are readily available (e.g., Refs. 44, 45), and the reader is assumed to be familiar with its application to continuous steel beams. These are often of uniform cross-section, so that the positive and negative moments of resistance, M_p and M_p', are equal. For the composite beam analysed in Chapter 3, $M_p = 770$ kNm, but the plastic moment M_{ps}, of the steel section alone is only 403 kNm. It is rarely practicable and never economic to provide enough slab reinforcement to make M'_p equal to M_p, so that the first step in a plastic analysis is to assume a preliminary value of μ, the ratio of M'_p to M_p. In beams for buildings, μ is likely to be between 0·5 and 0·7.

Two simple examples of the method are now given. Both relate to the design of a beam of span L for a distributed ultimate load w per unit length.

If the beam is continuous at both ends, Fig. 4.4(a), hinges occur at the ends and at midspan, and

$$(1 + \mu)M_p = wL^2/8 \tag{4.5}$$

If the beam is continuous at one end only, the bending moment diagram at collapse is as shown in Fig. 4.4(b). It can easily be shown that

$$\beta = \left(\frac{1}{\mu}\right)[(1 + \mu)^{1/2} - 1] \tag{4.6}$$

and

$$M_p = \tfrac{1}{2}w\beta^2 L^2 \tag{4.7}$$

There are certain conditions that must be satisfied for the plastic theory to be valid for composite beams. These are discussed in Section 4.4.

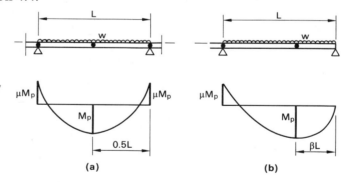

Fig. 4.4

4.3.3   Example (continued). Negative moment of resistance

In simple plastic theory, no account is taken of load history or of the loading on spans other than the one being designed, and the section properties are the same for dead as for imposed load. The relevant collapse mechanism for each span of the two-span beam being designed is that of Fig. 4.4(b), with $w = 70$ kN/m (Table 4.1) and $L = 10\cdot5$ m. As a first step, the beam section used in Chapter 3 is considered, with $A_r = 0$. So $F_r = 0$, and from (4.3) and Section 3.4.2, $M_p' = M_s = 403$ kNm. Then $\mu = 403/770 = 0\cdot52$.

Some slab reinforcement must be provided in the negative-moment region to control cracking, so μ will inevitably exceed $0\cdot52$. We next consider a beam with $\mu = 0\cdot6$.

From (4.6),

$$\beta = 1\cdot67 \times 0\cdot264 = 0\cdot44$$

From (4.7),

$$M_p = 35 \times 0\cdot44^2 \times 10\cdot5^2 = 751 \text{ kNm}$$

Therefore

$$M_p' = 0\cdot6 \times 751 = 451 \text{ kNm}$$

These are a possible set of plastic moments of resistance at midspan and the internal support.

The cross-section shown in Fig. 3.7 has $M_p = 770$ kNm, and so is satisfactory. The slab reinforcement at the internal support is now designed. To ensure that the stress at working load is not excessive, it is prudent to provide a margin over the 451 kNm calculated above. Equation (4.3) can be used for a preliminary calculation. We assume first that d_{wc} (Fig. 4.3(a)) is $d - h_c - t_f$, or $0\cdot19$ m in this case. Then

$$d_{sc} = d - \tfrac{1}{2}d_{wc} = 0\cdot356 - 0\cdot095 = 0\cdot261 \text{ m}$$

We now design for $M_p' = 480$ kNm, rather than 451 kNm, and assume $d_r = 0\cdot04$ m. With $M_{ps} = 403$ kNm, (4.3) gives

$$F_r = (480 - 403)/0\cdot221 = 348 \text{ kN}$$

If high-yield bars are used ($f_{rd} = 0\cdot87 \times 410 = 357$ N/mm^2), then from (4.1),

$$A_r = 348\,000/357 = 976 \text{ mm}^2$$

Six 16-mm bars have $A_r = 1206$ mm^2, and will be used. Then

$$F_r = 1206 \times 0\cdot357 = 430 \text{ kN} \tag{4.8}$$

They are placed below the transverse bars, so from Fig. 3.15,

$$d_r = 21 + 6 + 8 = 35 \text{ mm}$$

The negative moment of resistance is now calculated. From (4.2) with $t_w = 9\cdot4$ mm,

$$d_{wc} = 1206 \times 357/9\cdot4 \times 304 \times 2 = 75\cdot3 \text{ mm}$$

Now $h_c = 150$ mm, $t_f = 16$ mm, and $d = 356$ mm, so $d - h_c - t_f = 190$. This exceeds $75\cdot3$, so d_{wc} is correctly given above, and

$$d_{sc} = d - \tfrac{1}{2}d_{wc} = 356 - 38 = 318 \text{ mm}$$

From (4.3) and (4.8),

$$M'_p = 403 + (0\cdot318 - 0\cdot035)430 = 525 \text{ kNm} \qquad (4.9)$$

The margin that this value gives above the 451 kNm calculated by plastic theory may seem large; but it is only about 8% of $wL^2/8$ for the span, and later calculations will show that it is not enough when unpropped construction is used.

4.3.4   Vertical shear

Both Parts of CP 117 state that vertical shear shall be assumed to be resisted by the steel section alone. This is obviously correct at an end of a simply supported beam, where it may not be possible to develop much composite action, but the situation in a continuous beam is more complex.

The well-known distribution of vertical shear stress in a reinforced concrete beam (Fig. 4.5(a)) is derived by dubious extension of the elastic theory; but it is obvious that the 'cracked' concrete between levels A and B does resist vertical shear, even if the stress distribution

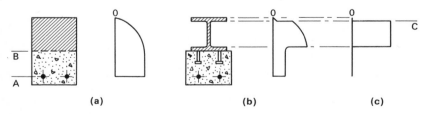

(a)   (b)   (c)

Fig. 4.5

is more complex than that shown. If the compression zone is replaced by a steel joist, we have the present problem (but upside-down), and the shear-stress distribution is by analogy as sketched in Fig. 4.5(b). Calculations show[46] that the proportion of vertical shear resisted by the concrete slab increases with the ratio A_r/A_s, as would be expected, and can in heavily reinforced beams be as high as 35%.

The shear distribution implied by CP 117 (Fig. 4.5(c)) is thus on the safe side. It is particularly conservative in design on an elastic basis, when allowance is made for interaction between bending and shear by calculating an equivalent stress. It may then be found that the combination of a high bending stress with the maximum shear stress at level C (Fig. 4.5(c)) may govern the design of the member.

In a continuous beam, the designer needs to know the extent to which the moment of resistance is reduced by shear, rather than the ultimate shear strength when the bending moment is low. For example, it is recommended[48] for rigid-jointed steel frames that no reduction in the flexural strength of a beam need be made when the shear is less than $V_p/\sqrt{3}$, where V_p, the shear capacity of an I-section of web depth d_w (between centres of flanges) and thickness t_w is given by

$$V_p = d_w t_w f_y/\sqrt{3} \qquad (4.10)$$

For shears exceeding $V_p/\sqrt{3}$, the available flexural strength is less than the moment of resistance M_{ps}, and drops to zero when V reaches V_p, as shown by the interaction curve in Fig. 4.6.

In research aimed at finding a similar ultimate-load method for composite beams, negative moment regions were subjected to very high shear forces.[46,47] It was found that the force ratio, Φ, given by

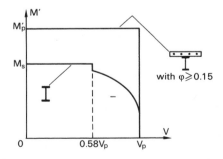

Fig. 4.6

$$\Phi = A_r f_{ry}/A_s f_y \tag{4.11}$$

is a more appropriate variable than A_r/A_s. Values of Φ used in the test specimens ranged from 0·15 to 0·69. The research led to a procedure for constructing curves that gave the negative moment of resistance M' as a function of the vertical shear at the cross-section considered. For design purposes, the results can be simplified by assuming that when the force ratio exceeds 0·15 and the steel member is not susceptible to buckling (as discussed in Section 4.4.2), the full plastic moment of resistance of the member, given by (4.3), can be developed when the vertical shear does not exceed V_p, as given by (4.10). This relationship is compared with that for a steel I-section in Fig. 4.6, which shows that longitudinal reinforcement improves resistance to shear as well as flexural strength.

4.3.5   Example (continued). Vertical shear

The beam is stronger than is required by plastic theory, so the moments of resistance (-525 kNm and $+770$ kNm) cannot both be reached at the design load of 70 kN/m. The bending-moment diagram at this load must lie between lines AB and CD on Fig. 4.7, but its exact position depends on the loading on the other span. The vertical shear at E should be taken as the maximum possible value, which occurs when the bending moment at E is -525 kNm. Then

$$V = \tfrac{1}{2} \times 70 \times 10·5 + 525/10·5 = 368 + 50 = 418 \text{ kN}$$

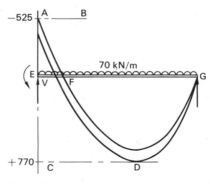

Fig. 4.7

From (4.11), the force ratio is

$$\Phi = 12{\cdot}06 \times 357/85{\cdot}3 \times 304 = 0{\cdot}166$$

which exceeds $0{\cdot}15$; and from (4.10),

$$V_p = (412 - 16) \times 9{\cdot}4 \times 0{\cdot}304/\sqrt{3} = 653 \text{ kN}$$

which exceeds 418 kN. Thus the vertical shear is well below the shear strength of the member. Web buckling is considered in Section 4.4.2.

4.3.6   Longitudinal shear

Half of a negative-moment region of a continuous beam with top longitudinal reinforcement of area A_r is shown in elevation in Fig. 4.8. Tests on beams have shown that at maximum load the slab near an internal support is severely cracked throughout its thickness; so the tensile force in the slab on line DF can be assumed to be F_r (eq. (4.1)). The force across BE, the line of contraflexure, is zero, so the shear connectors in length L_n have to resist a force F_r. For reasons similar to those given in Section 3.5,[49] these connectors can be spaced uniformly along EF.

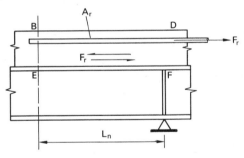

Fig. 4.8

As explained in Section 2.5, it has been recommended[22] that the strength of connectors in negative-moment regions should be taken as 20% less than in positive-moment regions. To avoid the confusing use of different design strengths, this can be implemented by giving the required number of connectors as

$$N = 1{\cdot}25F_r/P_d \qquad (4.12)$$

where P_d is $0.8P_k$, as explained in Section 3.5. The pitch or spacing of single connectors is then

$$p = L_n/N \tag{4.13}$$

It is proposed that the rules for transverse reinforcement should be the same as for positive-moment regions (Section 3.6). The research on this subject[26, 27] did not show any need for additional steel, even though the cracking of concrete might be expected to reduce its ability to transfer shear. This rule is simpler than that given in CP 117: Part 2, in which the amount of transverse reinforcement is related to the maximum tensile stress in the concrete slab at the cross-section considered. The proposal does in fact imply the use of more transverse reinforcement for a given shear than in a midspan region, because the design rules are related not to the calculated shear flow but to the number of connectors provided, which is higher for a given shear due to the factor 1.25 in (4.12). In this way some allowance is made for the biaxial shear that occurs in the slab when a two-way system of composite beams is used.

4.3.7   Example (continued). Shear connection and transverse reinforcement

When the negative moment at E (Fig. 4.7) is 525 kNm, the length EF is 1.43 m. From (4.8), the force F_r is 430 kN. Assuming that 100 mm $\times$ 19 mm headed studs are used, as before, $P_d = 69.5$ kN from Section 3.5.1. From (4.12),

$$N = 1.25 \times 430/69.5 = 7.7$$

say eight, and from (4.13),

$$p = 1.43/8 = 0.178 \text{ m}$$

This is for single studs; so four pairs of studs will be provided at 0.35 m pitch. This is a wider spacing than in the positive moment region, where 0.24 m was used, because of the relatively low force ratio.

The shear force per unit length of beam transferred to the slab is also lower than that in the positive-moment region $(430/1.43 = 300$ kN/m cf. $2600/4.5 = 578$ kN/m). A negative-moment region is so small a proportion of the span of a beam that in this situation one would provide the same transverse reinforcement as in the midspan region (Section 3.6.7).

4.4 The applicability of simple plastic theory

The ease with which plastic collapse mechanisms can be found and the simplicity of results such as (4.5) and (4.7) make the use of plastic theory attractive to designers. Its validity was established initially by means of research on rigid-jointed steel structures with members of solid rectangular or compact I-section. Later work showed that it could not be used for members with slender webs or flanges, nor for columns under high axial load, because of problems of instability. The theory gives safe results for most steel members because stress–strain curves for steel show a long yield plateau followed by strain hardening.

It has not been found possible to use it for reinforced concrete beams because the stress–strain curve for concrete in compression reaches its peak at a strain less than one-twentieth of that reached by steel during strain-hardening, and then drops sharply (Fig. 4.9).

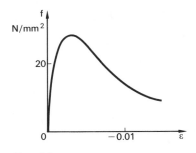

Fig. 4.9

The properties of composite members are a combination of those of steel and of concrete, so the question of applicability must be answered. The theory is 'applicable' if it gives results that are 'on the safe side', but not excessively so. To be more specific: designers need assurance that if a beam like that of the present example but with calculated plastic moments of resistance $M_p = 751$ kNm, $M'_p = 451$ kNm (as found in Section 4.3.3) were constructed and loaded to failure, the total load for one span at failure would be not less than 70 kN/m, whether the other span were loaded or not, and whether propped or unpropped construction were used. It will be shown that this assurance can be given for beams of compact cross-section.

Another question is whether the designer may assume without further check that the crack widths and deflections of the beam will satisfy the requirements of the serviceability limit state. This subject is considered in Sections 4.5 and 4.6.

4.4.1   Moment redistribution in a fixed-ended beam

The transition from elastic to plastic behaviour in a continuous beam under increasing load involves redistribution of longitudinal bending moments, to an extent that is greater in a composite beam than in a steel beam. To illustrate the typical behaviour of composite beams, a step-by-step analysis is given in Appendix B of the response of a fixed-ended beam of span L to increasing load w per unit length. The assumed moment–curvature relation for the negative-moment region (Fig. B.2) reproduces the low rotation capacity (due to buckling of the steel member) that causes slender beams to fail at loads lower than that given by simple plastic theory.

The plastic theory predicts the three-hinge collapse mechanism shown in Fig. 4.4(a). For a beam with $M_p = 2M'_p$, the collapse load w_p is given by (4.5) as

$$w_p = 8(1 + 0.5)M_p/L^2 = 12M_p/L^2 \qquad (4.14)$$

In the beam studied in Appendix B, yield first occurs at the supports at a load of $0.375w_p$, and the hinges at these points are fully formed when $w = 0.546w_p$. At $w = 0.76w_p$, the rotation capacity of the negative-moment region is exhausted, when the midspan moment is still only $0.64M_p$. Further deformation and the formation of the midspan hinge occur under diminishing load; so the beam fails at three-quarters of the load given by simple plastic theory.

These results are typical of the behaviour of composite beams in which negative moment regions have low rotation capacity. The negative hinges always form first in fixed-ended beams, and usually do so in other beams, so the shape of the moment-curvature curves for regions near supports has far more influence on ultimate strength than does the behaviour of the midspan region. In composite beams generally, the degree of redistribution of elastic moments that must occur after first yield is greater than in steel beams. For these reasons, simple plastic theory can only be used safely for beams in which large hinge rotations can occur in negative moment regions without loss of flexural strength.

Thus it became evident that the criteria of suitability for plastic design must be based on a study of moment–rotation characteristics of negative moment regions. This work is now described.

4.4.2   Local buckling in negative-moment regions

When the steel member in the negative-moment region of an uncased composite beam is of I-section, it invariably begins to buckle before the maximum negative moment is reached (Plate 6). Van Dalen[24] tested isolated negative-moment regions with stocky cross-sections, and found that buckling began at an extreme fibre compressive strain of

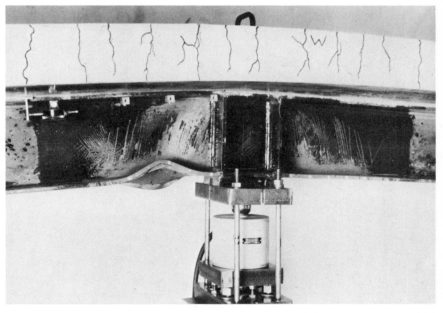

Plate 6. Local buckling of the compression flange of a continuous composite beam near an internal support (courtesy Dr J. Climenhaga)

about 0·017. He defined M'_p as the negative moment when this strain was first reached, and assumed that this corresponded to the maximum load that the beam could carry, w_{max}. Using moment-curvature curves obtained in earlier research[50] he computed w_{max}/w_p for seventy fixed-ended beams of various proportions, with force ratios Φ ranging from 0 to 0·8, and found that the results all lay within the shaded band

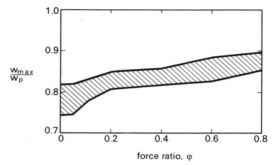

Fig. 4.10

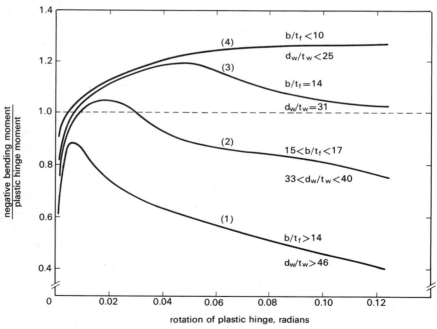

Fig. 4.11

in Fig. 4.10. But in tests on continuous beams, w_{max}/w_p was usually found to exceed 1·0, so Climenhaga made a more detailed study of moment–rotation curves. Figure 4.11, based on his test results[51,52] shows that there are four basic shapes of curve, depending on the web and flange slendernesses of the steel member. As may be expected from the worked example in Appendix B, it was concluded that the members that gave curves of types (1) or (2) were unsuitable for plastic design.

This work led to the following limiting slenderness ratios for the webs and compression flanges of unstiffened steel members suitable for design by simple plastic theory. They depend on the yield stress of the steel, f_y, which can be expressed nondimensionally in terms of the yield strain ε_y, where

$$\varepsilon_y = f_y/E_s.\qquad(4.15)$$

Other symbols are shown in Fig. 4.12.

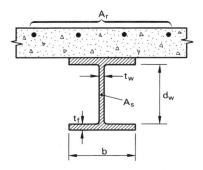

Fig. 4.12

For compression flanges,

$$b/t_f \leqslant \frac{0·70}{[\varepsilon_y(3·18 - f_u/f_y)]^{1/2}}\qquad(4.16)$$

where f_u/f_y is the specified minimum ratio of tensile strength to yield strength for the steel considered. For steels to BS 4360:1972,[53] (4.16) gives:

$$\left.\begin{array}{l} b/t_f \leqslant 16·8 \text{ for Grade 43 steel} \\ b/t_f \leqslant 12·4 \text{ for Grade 50 steel} \end{array}\right\}\qquad(4.17)$$

It is shown in Section 4.3.1 that at M_p', the presence of top longitudinal reinforcement A_r has the effect of imposing a net compressive force $A_r f_{rd}$ on the steel member. This has little effect on the stress in the compression flange, but increases the mean compressive stress in the web, and so makes it more likely to buckle. This influence can be represented by the force ratio Φ, and led to these slenderness rules for webs:

$$\begin{array}{ll} d_w/t_w \leqslant 2\cdot44(1 - 1\cdot4\Phi)/\varepsilon_y^{1/2} & 0 \leqslant \Phi \leqslant 0\cdot28 \\ d_w/t_w \leqslant 1\cdot48/\varepsilon_y^{1/2} & \Phi > 0\cdot28 \end{array} \right\} \quad (4.18)$$

For steels to BS 4360, (4.18) gives

for Grade 43 $d_w/t_w \leqslant 70(1 - 1\cdot4\Phi)$ with a lower limit of 43
for Grade 50 $d_w/t_w \leqslant 58(1 - 1\cdot4\Phi)$ with a lower limit of 36 $\left.\right\}$ (4.19)

When $\Phi = 0$, the effective cross-section in the negative-moment region is simply a steel I-section. Equation (4.19) then gives limits, 70 and 58, which are well below the current British limits[48] of 85 and 70, respectively, for plastic design of rolled I-sections of steel to Grades 43 and 50. The reasons for this lack of agreement are

(1) that the existing rules ensure that M_p' can be reached, but not that it can be maintained over a hinge rotation large enough to ensure hinge formation at midspan and hence the attainment of w_p, the load given by simple plastic theory; and
(2) that the rotation required in a composite beam is more than in a steel beam.

The current rules for steel beams are therefore not applicable to composite beams.

4.4.3   Strain-hardening and its effects

There are several reasons why tests on beams satisfying these rules show that $w_{max}/w_p > 1$, even though the result of the example quoted above was only $0\cdot76$. Comparison of Fig. B.2 with the first and second curves in Fig. 4.11 shows that real beams have much larger rotation capacity, and that due to strain-hardening, the maximum negative moment of resistance exceeds M_p' by an average of 20%. The worked example did not allow for this important aspect of beam behaviour.

When a real fixed-ended beam with $M_p = 2M_p'$ is loaded to w_p, the end and midspan moments may be not M_p' and M_p, as the designer assumes, but $1{\cdot}3M_p'$ and $0{\cdot}85M_p$. For distributed load, this change increases the length of each negative-moment region from $0{\cdot}09\,L$ to $0{\cdot}125L$. This need not influence the detailing of the shear connectors, because the total number required in each half span is not altered, but it increases the length of longitudinal reinforcement in tension, and so makes it essential that these bars be given generous anchorage beyond the calculated points of contraflexure.

4.4.4   Rotation capacity of positive-moment regions

It is not sufficient to establish that plastic theory works for fixed-ended beams, which rarely occur in practice. Real beams have unequal spans, uneven loads, and various degrees of end fixity. Hope Gill has studied problems of rotation capacity in such beams (Plate 7), and has shown[54] that sometimes the midspan hinge forms before those at the supports. Calculations using moment–curvature curves based on standard stress–strain curves for steel and concrete again show that w_{max}/w_p can be less than one; and tests on beams have again shown w_{max}/w_p to exceed one. Agreement between theory and tests has only

Plate 7. Test on a continuous composite beam to study rotation capacity and the applicability of simple plastic theory (courtesy Dr M. Hope-Gill)

been obtained by allowing in the theory for the fact that, due to reinforcement and lateral restraint, concrete in compression in a wide reinforced slab can reach a higher compressive strain without serious loss of strength than it can in a small unreinforced specimen. The greater curvature allows more strain-hardening to occur in the tension flange of the steel member, so that the maximum positive moment exceeds M_p. This compensates for the deficiency in rotation capacity and allows w_p to be reached—but again not in the way the designer assumes.

4.4.5   Example (continued). Applicability of plastic theory

It is necessary to check that the steel I-section satisfies the limiting slendernesses given in Section 4.4.2. The steel is Grade 50, and from Fig. 3.7, the relevant dimensions of the cross-section are $b_f = 153$ mm, $t_f = 16$ mm, $d_w = 412 - 32 = 380$ mm, $t_w = 9.4$ mm; so that

$$b_f/t_f = 9.6 \qquad d_w/t_w = 40.4$$

From p. 113, the force ratio is 0·166, so from (4.17) and (4.19), the limiting slendernesses are

$$b_f/t_f \leqslant 12.4 \qquad d_w/t_w \leqslant 58(1 - 1.4 \times 0.166) = 44.5$$

These conditions are satisfied, so it may be assumed that the design method used in Section 4.3.3 is applicable to this beam.

4.5   Elastic analysis for stresses and deflections in service

A more detailed account is now given of the problems associated with elastic analysis of continuous composite beams that were outlined in Section 4.1. The two-span beam studied in previous sections will be used to illustrate procedures appropriate to beams in buildings. Beams in bridges are considered in Volume 2.

4.5.1   Properties of cross-sections subjected to negative moments

For a given modular ratio, there are six values for the second moment of area for a composite beam, three of which are useful in practice.

The concrete slab may be assumed to be uncracked (U), cracked due to positive moment (C+), or cracked due to negative moment (C−); and for each of these three alternatives, the slab reinforcement within the effective breadth may be included (I) or neglected (N). For brevity, each value of the second moment of area is now represented by the symbols for the assumptions made in calculating it; e.g., (U, I) is the value for the *Uncracked* section when the reinforcement is *Included*.

The designer often needs to estimate flexural rigidity before the slab reinforcement has been designed; and when the slab is in compression, its inclusion makes little difference to the result. So (U, I) and (C + , I) are rarely used. The value (C −, N) is the second moment of area of the steel section. It is easily calculated or found in tables, but has little relevance to a continuous beam unless the force ratio is very low.

In positive-moment regions, (C+, N) is used for calculation of stresses. The area of cracked concrete is usually small, so (U, N), which is easily calculated, differs little from (C+, N). Either value can be used when an average flexural rigidity for the whole span is required, as in the analysis of rigid-jointed frames. In negative-moment regions, (C−, I) is needed for the calculation of stresses. Formulae for (C+, N) are given in Section 3.7.1, and those for (U, N) and (C−, I) are now discussed.

Equations (3.28) and (3.30), derived for (C+, N) when the neutral axis lies in the steel, relate to an uncracked section, and so give the value (U, N) for all positions of the neutral axis.

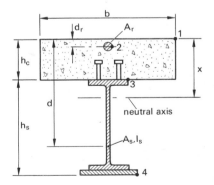

Fig. 4.13

When the concrete is cracked in negative bending and the slab reinforcement A_r is included, combination $(C-, I)$, the neutral axis invariably lies within the steel member. Its position is given by

$$A_r(x - d_r) = A_s(d - x) \qquad (4.20)$$

and the second moment of area (in 'steel' units) by

$$I = I_s + A_s(d - x)^2 + A_r(x - d_r)^2 \qquad (4.21)$$

The notation is shown in Fig. 4.13, which also defines levels 1 to 4 in the cross-section. Section moduli are required for concrete at level 1 and for steel at levels 2, 3, and 4. Assuming sagging moments and tensile stress to be positive, these are:

$$\left.\begin{array}{ll} Z_{c1} = -mI/x & \text{with } I \text{ given by } (3.30) \\ Z_{s2} = -I/(x - d_r) & \text{with } I \text{ given by } (4.21) \\ Z_{s3} = -I/(x - h_c) \ \big\} & \text{with } I \text{ given by } (3.30) \text{ or} \\ Z_{s4} = I/(h_c + h_s - x) \big\} & (4.21) \text{ as appropriate} \end{array}\right\} \quad (4.22)$$

4.5.2   Example (continued). Elastic properties of the cross-section

The numerical values given in Section 3.7.2 are relevant, and from Section 4.3.3, $A_r = 12 \cdot 06$ cm^2 and $d_r = 35$ mm. The results for x and I are given in Table 4.2; those marked * are taken from Section 3.7.2.

For case (U, N) with $m = 7 \cdot 5$, $x = 12 \cdot 9$ cm from (3.28); and from (3.30), $I = 85\,000$ cm^4. The inclusion of the reinforcement, case (U, I), gives $I = 85\,900$ cm^4, and so makes little difference.

For case $(C-, I)$, from (4.20), $x = 31 \cdot 6$ cm, and from (4.21), $I = 34\,690$ cm^4, for both values of m.

Table 4.2. Neutral axis depth and second moment of area of composite cross-section

Assumptions	$m = 7 \cdot 5$		$m = 15$	
	x cm	I cm^4	x cm	I cm^4
Cracked +, unreinforced $(C+, N)$	12·7	84 920	16·5	72 870*
Uncracked, unreinforced (U, N)	12·9	85 000	16·5	72 870*
Cracked −, reinforced $(C-, I)$	31·6	34 690	31·6	34 690

Section moduli for the midspan region are given in Table 4.3, and those for the negative-moment region, calculated from (4.22), are given in Table 4.4. For the case (U, N) it is seen that the effect of modular ratio on Z_{c1} is significant, and greater than its effect on I. In the midspan region, only the stress at level 4 is likely to be important in practice, and Table 4.3 shows that the effect of modular ratio on this stress is small.

Table 4.3. Section moduli for positive moments, based on the cracked unreinforced cross-section (C+, N)

Section modulus	Values of Z (cm³)	
	$m = 7.5$	$m = 15$
Z_{c1}	− 49 990	− 66 250
Z_{s3}	37 580	− 48 580
Z_{s4}	1 954	1 836

Table 4.4. Section moduli of the composite section for negative moments

Section modulus	Relevant I	Values of Z (cm³)	
		$m = 7.5$	$m = 15$
Z_{c1}	U, N	− 49 500	− 66 250
Z_{s2}	C−, I		− 1230
Z_{s3}	C−, I		− 2090
Z_{s4}	C−, I		1410

4.5.3 Longitudinal bending moments and stresses

The designer of a composite beam must know whether propped or unpropped construction is to be used, for the stresses and deflections of the beam in service are much influenced by this decision. But in buildings it is uneconomic to specify a construction sequence and timetable in detail, and impracticable to ensure that it is followed, even though this also affects stresses in service. It must therefore be accepted that these cannot be calculated accurately. Their values have more significance in bridges than in buildings, so this subject is discussed further in Volume 2.

Loading.   When design of beams for buildings is based on unpropped construction, it is assumed that the steel beam carries itself and the weight of the concrete floor slab, and that all other loading is carried by the composite member. The highest stresses at internal supports occur when imposed loading acts only on the two adjacent spans; but it is accurate enough when calculating these stresses to assume that all spans are fully loaded. Stresses at midspan are highest when only that span is loaded. These stresses rarely govern design of continuous beams, so that the relevant calculations can usually be omitted.

When propped construction is specified, it is assumed in design that all load is carried by the composite member. This implies that props must remain under each span until the concrete in the slabs for that span *and adjacent spans* has reached at least three-quarters of the specified characteristic cube strength.

Relative stiffness of members.   For the calculation of the longitudinal moments, the ratios of the stiffnesses of the members are required. In reinforced concrete members, these are usually based on the gross (uncracked) cross-section of the concrete, and so are independent of the modular ratio. Thus the effects of cracking and of creep do not influence the analysis of the structure. The development of an equally simple method of analysis for composite structures is now discussed.

The cracking of concrete has more influence on the distribution of longitudinal moments than in a reinforced concrete beam, because it occurs mainly in negative-moment regions, and not at midspan unless the steel section is encased. In an accurate analysis for longitudinal moments it would therefore have to be assumed that the cross-section of every beam varies along its length. This is too complicated for everyday use, so we must accept that in practice, relative stiffnesses of members will be based on uncracked sections, as in reinforced concrete. In studying the consequences of this, it must be remembered that analysis at the serviceability limit state has two objectives only: to ensure that deflections and vibrations are not excessive; and, where a concrete surface is exposed to view or to a corrosive environment, to determine the spacing of reinforcement bars needed to control the width of cracks. There is no need to check strength, for that is done at the ultimate limit state, and it does not matter if calculations show that steel has yielded (unless members are slender), provided that allowance is made for the effect of this on deflections and crack widths.

Redistribution of moments.   The influence of cracking of concrete near internal supports of continuous beams on the distribution of longitudinal moments was extensively studied during the development of draft clauses for the design of composite bridges.[15] For beams of the proportions used in buildings, the bending moment at an internal support at the serviceability load is likely to be between 15 and 30% lower than that given by uniform-section elastic analysis.

It is accurate enough to assume that the results of such an analysis can be corrected to allow for cracking simply by redistributing 20% of the moment carried by the composite section at each internal support, and making corresponding increases in adjacent midspan moments.

Local yielding at the support may cause further redistribution of moments. It is proposed that this be allowed, up to a further 20% of the support moment given by the initial analysis.

In practice, it is convenient to do both redistributions at once, by calculating the stresses in the steel at the supports for the moments given by elastic theory, and then redistributing these moments by an amount (not exceeding 40%) that is just sufficient to reduce all the steel stresses to or below the limiting stresses specified at the serviceability limit state. This method is illustrated in Section 4.5.4, in which the limiting stresses are taken as $1 \cdot 0 f_y$ and $1 \cdot 0 f_{ry}$.

It is possible to estimate the degree of redistribution due to cracking by a method based on the draft proposals for the design of bridge beams, as follows. At each internal support, the tensile stress at the top surface of the concrete slab due to the peak negative moment is calculated, assuming an *uncracked* cross-section. The elastic properties of the beam can usually be assumed for this purpose to be the same as at midspan. Let this stress be f_{c1}. The appropriate degree of redistribution from that support into adjacent spans is $80 f_{c1}/f_{cu}\%$, where f_{cu} is the characteristic cube strength.

In unpropped construction, the bottom-fibre stress in the steel beam at midspan after redistribution should be checked, and the design revised if this is found to exceed yield. In propped construction, this stress is usually found to be well below yield. As in the design of reinforced concrete, there is no need to check the compressive stress in the concrete slab at the serviceability limit state.

Modular ratio.   When calculating stresses by elastic theory, a decision has to be made about the extent to which allowance is made for the effects of creep of concrete. As explained in Section 3.7.1, this is done by using an increased value of modular ratio. Both CP 117: Part 2 and Appendix A of CP 110 recommend that the value for short-term loading be doubled for sustained load; that is, that the effective modulus E'_c be taken as $\frac{1}{2}E_c$. Design for serviceability is influenced much less by stresses in concrete than by stresses in steel, which are increased by creep, so it is on the safe side to assume $E'_c = \frac{1}{2}E_c$ when calculating stresses due to all loading. This useful simplification is likely to be over-conservative only in the rare situation where un-propped construction is used and almost the whole load carried by the composite member is 'short-term'. Then the assumption $E'_c = E_c$ would be more appropriate.

4.5.4   Example. Stresses due to serviceability loads

The beam to be analysed has two spans each of 10·5 m (Fig. 4.1). As explained above, its cross-section will initially be assumed to be uniform along its length. The relative stiffness of the spans is then independent of modular ratio. The following results given by elastic theory for such a beam with two spans L will be used.
For a distributed load w on one span only:

$$\text{at the internal support} \quad M' = -0\cdot0625wL^2 \quad (4.23)$$
$$\text{in the loaded span} \quad M_{\text{max}} = 0\cdot096wL^2 \quad (4.24)$$

For distributed load w on both spans:

$$\text{at the internal support} \quad M' = -0\cdot125wL^2 \quad (4.25)$$
$$\text{in the loaded spans} \quad M_{\text{max}} = 0\cdot070wL^2 \quad (4.26)$$

From Table 4.1, the distributed loads at the serviceability limit state are

$$\text{dead load} \quad g = 15 \text{ kN/m} \quad (4.27)$$
$$\text{imposed load} \quad q = 30 \text{ kN/m} \quad (4.28)$$

Unpropped construction.   It is assumed that the whole of the dead load is carried by the steel section. From (4.25) and (4.27), the negative moment at the internal support when both spans are loaded is

$$M'_g = -0{\cdot}125 \times 15 \times 10{\cdot}5^2 = -207 \text{ kNm} \qquad (4.29)$$

The section modulus of the steel member is $23\,800/20{\cdot}6 = 1155 \text{ cm}^3$ for both flanges, so the dead-load stresses at levels 3 and 4 (Fig. 4.13) are

$$f_{s3} = f_{s4} = \pm\, 207/1{\cdot}155 = \pm\, 179 \text{ N/mm}^2 \qquad (4.30)$$

The stress distribution is shown in Fig. 4.14. If the yield stress of the steel (350 N/mm^2) is not to be exceeded, the stress f_{s4} due to load carried by the composite member must not exceed $350 - 179 = 171$ N/mm^2.

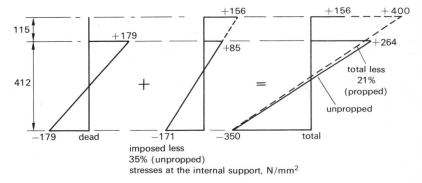

Fig. 4.14

When both spans are loaded, the support moment due to imposed load is

$$M'_q = -0{\cdot}125 \times 30 \times 10{\cdot}5^2 = -414 \text{ kNm} \qquad (4.31)$$

Using the section moduli given in Table 4.4, the stresses in the steel due to this moment are

$$f_{s2} = +\,414/1{\cdot}23 = +\,336 \text{ N/mm}^2$$
$$f_{s4} = -\,414/1{\cdot}41 = -\,294 \text{ N/mm}^2$$

The stress f_{s4} is the more critical. To bring it down to 171 N/mm^2 would require a redistribution of imposed-load moment of $(294 - 171)/294 = 0{\cdot}42$, or 42%. It would therefore be prudent to strengthen the cross-section.

The simplest alteration that can be made is to increase the area of top longitudinal reinforcement, A_r. This is not a very effective method of reducing the stress f_{s4}, as the following figures show, nor is it needed for the control of crack width (p. 140).

If twelve 16-mm bars are used, A_r is increased from 12·06 cm² to 24·1 cm². The revised properties of the cross-section are found by the method of Section 4.5.1 to be

$$x = 286 \text{ mm} \qquad I(C-, I) = 43\,180 \text{ cm}^4$$
$$Z_{s2} = 1720 \text{ cm}^3 \qquad Z_{s4} = 1565 \text{ cm}^3$$

The new stress f_{s4} due to imposed load is -264 N/mm², and the amount of redistribution now needed is 35%. The new stress in the slab reinforcement is $+240$ N/mm², reduced by redistribution to 156 N/mm², so that this material, which has a yield stress of 410 N/mm², is not being used efficiently.

The most effective method of strengthening the beam would be to add a short bottom-flange cover plate near the internal support; but in practice a slightly heavier steel section would probably be used for the whole span.

The maximum stresses in the midspan region are now calculated, assuming that the amount of top reinforcement at the support has been increased to 24·1 cm², that 35% of the imposed-load moment at the support is redistributed, and that the increase in midspan moment due to redistribution is half the reduction in support moment. A modular ratio of 15 is used, and the relevant section moduli are given in Table 4.3. The stress distributions are shown in Fig. 4.15. Dead-load moment (both spans loaded)

$$M_g = 0{\cdot}07 \times 15 \times 10{\cdot}5^2 = 116 \text{ kNm} \qquad (4.32)$$

Stresses in steel

$$f_{s3} = f_{s4} = \mp\, 116/1{\cdot}155 = \mp\, 100 \text{ N/mm}^2 \qquad (4.33)$$

Imposed-load moment (both spans loaded) with redistribution:

$$M_q = 0{\cdot}07 \times 30 \times 10{\cdot}5^2 + \tfrac{1}{2} \times 0{\cdot}35 \times 414 = 304 \text{ kNm}$$

Imposed-load moment (one span loaded):

$$M_q = 0{\cdot}096 \times 30 \times 10{\cdot}5^2 = 317 \text{ kNm} \qquad (4.34)$$

The higher value (317 kNm) will be used. It gives the following stresses in the composite section:

$$f_{c1} = - 317/66\cdot2 = 4\cdot8 \text{ N/mm}^2 \qquad (16\% \text{ of the cube strength})$$
$$f_{s4} = 317/1\cdot84 = 173 \text{ N/mm}^2$$

From (4.33), the total stress f_{s4} is 273 N/mm² (78% of the yield stress). Thus even with 35% redistribution, midspan stresses are quite low, particularly in the concrete slab.

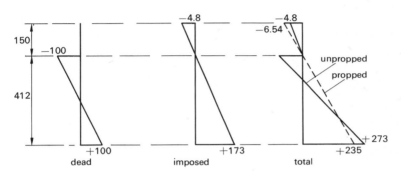

stresses at midspan, N/mm²

Fig. 4.15

Propped construction. The whole load is carried by the composite section, and the moments before redistribution are as follows:
at the internal support, from (4.29) and (4.31),

$$M' = - 207 - 414 = - 621 \text{ kNm} \qquad (4.35)$$

at midspan, for imposed load on both spans, from (4.32)

$$M = 3 \times 116 = 348 \text{ kNm} \qquad (4.36)$$

and for imposed load on one span, from (4.34),

$$M = 116 + 317 = 433 \text{ kNm} \qquad (4.37)$$

The extreme-fibre stresses at the support are

$$f_{s2} = + 621/1\cdot23 = + 505 \text{ N/mm}^2 \qquad (= 1\cdot23 f_{ry})$$
$$f_{s4} = - 621/1\cdot41 = - 440 \text{ N/mm}^2 \qquad (= 1\cdot26 f_y)$$

To reduce f_{s4} to the yield stress, 21% of the support moment is re-distributed, increasing the midspan moment by $\frac{1}{2} \times 0.21 \times 621 = 65$ kNm. Adding this to result (4.36) gives 413 kNm, so the value due to imposed load on one span only (433 kNm) still governs.

The stresses at the support are now

$$\left.\begin{array}{l} f_{s2} = 0.79 \times 505 = 400 \text{ N/mm}^2 \\ f_{s4} = -0.79 \times 440 = 350 \text{ N/mm}^2 \end{array}\right\} \qquad (4.38)$$

and those at midspan (with $m = 15$) are

$$f_{c1} = -433/66.2 = -6.54 \text{ N/mm}^2 \qquad (22\% \text{ of the cube strength})$$
$$f_{s4} = +433/1.84 = +235 \text{ N/mm}^2 \qquad (67\% \text{ of the yield stress})$$

The relatively low amount of redistribution needed and the low midspan stresses suggest that the beam as originally designed (with $A_r = 12.06 \text{ cm}^2$) would be satisfactory when propped construction is used; but a conclusion on this point is deferred until deflections and crack widths have been examined, in Sections 4.5.5 and 4.6.

Both for unpropped and propped construction, the figures show clearly the importance of checking stresses at the internal support, and that midspan stresses in steel in propped construction and in concrete generally are so low that normally they need not be checked.

To provide a check on the assumed amount of redistribution due to cracking (20%), estimates are now made in accordance with the method given on p. 127. The top-fibre section modulus for the un-cracked cross-section is given in Table 4.3 as 66 250 cm³ when $m = 15$. The top-fibre concrete stresses are

unpropped $f_{c1} = -414/66.2 = 6.27 \text{ N/mm}^2 = 0.21 f_{cu}$
propped $f_{c1} = -621/66.2 = 9.4 \text{ N/mm}^2 = 0.313 f_{cu}$

The values of $80 f_{c1}/f_{cu}$ are therefore 17% and 25%, so that 20% is a good approximation in this instance.

4.5.5   Deflections of continuous beams

Continuous composite beams are used mainly in long-span structures, where it is wise to check deflections because they may be large enough to cause problems with partitions or drainage unless appropriate precautions are taken.

As explained in Section 4.5.3, some redistribution of moments should be allowed at serviceability loads. Deflections may then exceed those calculated by elastic theory when beams are assumed to be of uniform flexural rigidity. A simpler and more accurate method is to consider each span separately, subjected to its design loading and the end moments after redistribution.

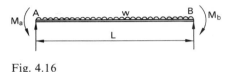

Fig. 4.16

We consider for example a span AB (Fig. 4.16) of a composite beam constructed 'propped' and subjected to distributed loading w and end moments M_a and M_b as shown. Unless one of the end moments is positive (i.e., tending to increase deflections), it is accurate enough to assume that the maximum deflection occurs at midspan. This deflection due to an end moment M is given by elastic theory as $ML^2/16EI$, so the midspan deflection of beam AB is

$$\delta = \frac{5wL^4}{384E_sI} - \frac{M_aL^2}{16E_sI} - \frac{M_bL^2}{16E_sI} =$$
$$\frac{L^2}{384E_sI}[5wL^2 - 24(M_a + M_b)] \quad (4.39)$$

where I refers to the composite section at midspan and is in 'steel' units. It is simplest to calculate I using an effective modulus E'_c that allows for creep, and to take w as the total load carried by the composite section. In unpropped construction, the deflection due to the load carried by the steel member alone must be calculated separately, with I in (4.39) replaced by I_s.

Limiting span–depth ratios[32] provide a simple method of controlling deflections in continuous beams, but are less accurate than the corresponding ratios for simply supported beams (Section 3.7.3).

4.5.6   Example. Deflections

The end moments calculated in Section 4.5.4 will be used. We consider only the right-hand span (Fig. 4.1), so M_b in Fig. 4.16 is taken as zero.

For propped construction, the end moment is given by (4.35) as

621 kNm. After redistribution,

$$M_a = 0.79 \times 621 = 490 \text{ kNm}$$

From Table 4.2 with $m = 15$, the relevant I is 72 870 cm⁴. Substituting $L = 10.5$ m, $E_s = 200 \text{ kN/mm}^2$, $w = 45 \text{ kN/m}$ in (4.39) gives

$$\delta = 1.98(24.8 - 11.8) = 25.8 \text{ mm} \qquad (4.40)$$

This is span/407. If M_a is taken instead as 621 kNm, the deflection is found to be 19.6 mm, so the effect of redistribution is significant here.

For unpropped construction, from (4.29) and (4.31), the end moments are

$$M_{ag} = 207 \text{ kNm} \qquad M_{aq} = 0.65 \times 414 = 269 \text{ kNm}$$

For dead load, $w = 15 \text{ kN/m}$, and $I_s = 23\,800 \text{ cm}^4$, so from (4.39),

$$\delta_g = 6.05(8.26 - 4.96) = 20.0 \text{ mm} \qquad (4.41)$$

For imposed load, $w = 30 \text{ kN/m}$ and $I = 72\,870 \text{ cm}^4$, so that

$$\delta_q = 1.98(16.5 - 10.1) = 19.9 \text{ mm} \qquad (4.42)$$

The total deflection, 39.9 mm, is span/264. It is likely to be excessive unless the steel beams are cambered. This example illustrates the need to check deflections, particularly in unpropped beams.

4.6   Crack-width control in continuous beams

In reinforced-concrete structures, one of the requirements for the serviceability limit state is that cracking of concrete shall not adversely affect the appearance or durability of the structure. In CP 110, 'reasonable limits' for maximum acceptable crack widths at the surface of a member are given as 0.3 mm generally, and 0.004 times the nominal cover to the main reinforcement for particularly aggressive environments. The recommended method of calculating crack widths, given in Appendix A of CP 110, is based on unpublished work for the European Committee for Concrete and on research by the Cement and Concrete Association.[31, 55] The spacing and width of cracks in nominally identical structures varies between wide limits, so sufficient testing was done to provide a sound statistical basis for the design method. The recommended limits are those which have a 20% probability of being exceeded when the full design load for the serviceability limit state acts on the structure.

In bridges, the main purpose of crack-width control is to ensure durability. It is then possible to neglect the short-term increase in crack widths that occurs when the full imposed load acts on the structure. For the draft Bridge Code, it has been proposed that crack widths should be calculated for the dead loading plus half the imposed loading, and that the design probability of excessive crack widths should be reduced from 20% to 5%. Similar arguments apply to certain types of building structure; but at present it seems necessary to develop crack-width rules for composite structures in buildings that are consistent with the method of CP 110.

In designing reinforced concrete structures, it is rarely necessary to calculate crack widths, for it has been found that satisfactory crack control can be obtained if the spacing of reinforcing bars does not exceed certain limits, calculated from the crack-width equations.[9] Similar bar spacing rules are being developed for use in concrete and composite bridges; but no rules are yet available for composite beams in buildings. A proposed method of crack-width control for negative-moment regions of such beams is now given, with a worked example.

There has as yet been no systematic research on cracking in composite beams, nor is it known to what extent the flat surfaces of encased steel sections can assist in the control of crack widths; so the present proposals are based solely on the research on reinforced concrete, and only the reinforcing bars are assumed to contribute to crack control.

The cracking at midspan in encased composite beams is considered in Section 4.6.2; but as encased beams are uncommon, the rules are developed first for cracking at the top surface of the concrete slab near an internal support for a continuous beam. This surface may be in the non-corrosive atmosphere of a heated building, and covered from view by a floor finish, and in this situation the designer may decide that no check on crack width is needed. But in a multi-storey car park with no floor finish, water containing de-icing salt may enter the cracks, and severe corrosion may result if their width is not controlled.

A cross-section of a negative-moment region of a composite beam is shown in Fig. 4.17(a). The longitudinal slab reinforcement consists of bars of diameter ϕ at pitch p with top cover c, and the total area of such steel within the effective breadth b is A_r. At the cross-section considered (which is usually that where the bending moment is greatest), M' is the negative moment calculated by 'uniform-EI' elastic theory due to the serviceability loads resisted by the composite section. It is reduced by redistribution (as explained in Section 4.5.3) to $\beta M'$. The stress in the reinforcement A_r due to $\beta M'$ and calculated for the elastic cracked cross-section is f_r. Due to redistribution, f_r will not exceed the design yield stress for the reinforcement. The neutral-axis depth is x (Fig. 4.17). This notation, used throughout this chapter, is inconsistent with that in CP 110, where x is defined as the neutral-axis depth measured from the extreme fibre in compression (i.e., x_c in Fig. 4.17).

In the present notation, the expression given in CP 110 for the design crack width w at a point on the surface of the concrete a distance a_{cr} from the surface of the nearest longitudinal reinforcing bar is

$$w = \frac{3a_{cr}\varepsilon_m}{1 + 2(a_{cr} - c)/x} \qquad (4.43)$$

where ε_m is the average strain at the level where cracking is being considered. It has been found that ε_m is less than the strain deduced from the calculated values of f_r and x (ε_1 in Fig. 4.17(b)) due to the stiffening effect of the concrete in tension between cracks. This alters

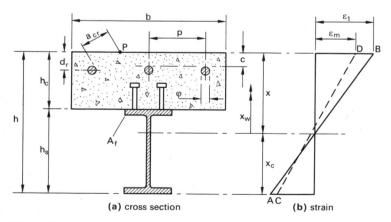

(a) cross section   **(b)** strain

Fig. 4.17

the distribution of mean longitudinal strain from AB in Fig. 4.17(b) to CD, and so reduces the surface crack width. The amount of this tension stiffening is given in the present notation by this equation from Appendix A of CP 110:

$$\varepsilon_m = \varepsilon_1 - \frac{1 \cdot 2bhx_w f_1}{A_{rt}xf_{yt}} \times 10^{-3} \qquad (4.44)$$

where x_w is the distance from the neutral axis to the point at which the crack width is being calculated, f_1 is a constant stress of 1 N/mm^2, re-expressed where necessary in units consistent with those used for f_{ry}, A_{rt} is the area of longitudinal reinforcement in tension, and f_{yt} is its yield stress.

If a very wet concrete mix is used, as it may be for the encasement of the tension flange of a composite beam, the free shrinkage of the concrete may exceed 0·0006. It is recommended in CP 110 that allowance should then be made for the effect of shrinkage on crack widths by adding half the expected shrinkage strain to ε_m. Values for free shrinkage may be estimated by a method given in the *Recommendations of the European Committee for Concrete*.[55]

Equation (4.44) appears in Ref. 9, but there seems to be no published derivation of it. It assumes that tension stiffening varies inversely with the proportion of tensile reinforcement (A_{rt}/bh) and with the stress in this steel, since this is assumed to be a constant proportion (0·58) of the yield stress f_{yt}. It agrees well with Beeby's equation[31] for tension stiffening when high-yield reinforcement is used, but the latter is more conservative for mild steel, since it assumes tension stiffening to be independent of the stress in the steel.

The applicability of (4.44) to negative-moment regions of composite beams is now considered. The equation is based on the assumption that the whole of the tension steel A_{rt} is close to the tension surface of the beam. In a composite beam, the top flange of the steel section (of area A_f, say) resists a significant proportion of the tensile force, so it is not obvious how A_{rt} should be defined when (4.44) is used for composite beams. The tension stiffening may be seriously over-estimated if we assume that A_{rt} equals A_r. It has been concluded from a theoretical study[56] that for most composite beams, a good approximation is $A_{rt} = A_r + 0·6A_f$; but until test data are available, a more conservative assumption is recommended:

$$A_{rt} = A_r + A_f \qquad (4.45)$$

Another problem is that it cannot be assumed in composite beams that the stress in the top reinforcement is $0.58f_{ry}$. Allowance can be made for this as follows. The calculations given in Section 4.5.3 lead to a known value of the stress in this steel (f_r) when the bending moment is $\beta M'$; and they include the approximation that the moment given by an elastic analysis allowing for cracking would be $0.8M'$. At this bending moment, the stress in the steel would be $0.8f_r/\beta$. (This result is accurate enough even if this stress exceeds yield, for tension stiffening varies with the strain in this steel, not the stress.) For (4.44) to be correct, this stress must be $0.58f_{yt}$, so that

$$f_{yt} = 0.8f_r/0.58\beta = 1.38f_r/\beta \qquad (4.46)$$

Substituting (4.45) and (4.46) in (4.44),

$$\varepsilon_m = \varepsilon_1 - \frac{0.9bhx_w\beta f_1}{(A_r + A_f)xf_r} \times 10^{-3} \qquad (4.47)$$

A similar adjustment must be made when calculating the strain ε_1. Assuming as above that the correct bending moment is $0.8M'$, then at level x_w in the member,

$$\varepsilon_1 = 0.8M'x_w/E_sI$$

But we know that $f_r = \beta M'(x - d_r)/I$. Eliminating M', we have

$$\varepsilon_1 = \frac{0.8x_wf_r}{E_s\beta(x - d_r)} \qquad (4.48)$$

The proposed procedure for checking crack width at the top of the slab is now summarised:

(1) From previous calculations, the dimensions of the cross-section and x, β, and f_r are known. Put $x_w = x$ and find ε_1 from (4.48).
(2) Calculate ε_m from (4.47). If it is negative, the slab is uncracked.
(3) Either an arrangement of reinforcing bars or a limiting crack width must now be assumed. The maximum crack width will occur midway between two bars, at a point such as P in Fig. 4.17 (a), and

the distance a_{cr} is then given by

$$d_r^2 + (\tfrac{1}{2}p)^2 = (a_{cr} + \tfrac{1}{2}\phi)^2 \tag{4.49}$$

(4) If d_r, p and ϕ are known, a_{cr} is found from (4.49) and then w from (4.43). Alternatively, (4.43) can be used to find a_{cr} for a given crack width w, and then p and ϕ can be chosen to satisfy (4.49).

Although few measurements have been made of crack widths in continuous composite beams, it has been observed in many tests that the cracking is generally similar to that in reinforced concrete T-beams, and it is believed that similar methods of crack control can be used. Allowance has been made in the method given above for the higher strains in top reinforcement that are shown by full-interaction theory to occur in composite beams; but no account has been taken of the effect of slip, which reduces tensile strains, and to this extent the method is conservative. Examples of its use are now given.

4.6.1   Example. Crack-width control

In the continuous beam shown in Fig. 4.1, the widest cracks may be expected to occur in the top surface of the concrete slab above the central support. The widths of these cracks will now be estimated, using the bending moments and stresses at the serviceability limit state calculated in Section 4.5.4, for slab reinforcement consisting of twelve 16-mm bars. The effective breadth of the slab is 1·8 m, so the bars are assumed to be placed at a uniform spacing of 0·15 m, with their centres 3·5 cm below the top of the slab. Therefore $c = 3\cdot5 - 0\cdot8 = 2\cdot7$ cm. From (4.49),

$$3\cdot5^2 + 7\cdot5^2 = (a_{cr} + 0\cdot8)^2$$

whence $a_{cr} = 7\cdot46$ cm.

The area of the top flange of the steel section is 24·8 cm²; other numerical values already known are $A_r = 24\cdot1$ cm², $h = 56\cdot2$ cm, $x = 28\cdot6$ cm, $b = 180$ cm, $E_s = 200$ kN/mm². For the top surface of the slab, $x_w = x = 28\cdot6$ cm.

From (4.43),

$$w = \frac{3 \times 74\cdot6\varepsilon_m}{1 + 2 \times 47\cdot6/286} = 168\varepsilon_m \quad \text{mm} \tag{4.50}$$

From (4.47),

$$\varepsilon_m = \varepsilon_1 - \frac{0.9 \times 180 \times 56.2 \times 10^{-3}\beta}{48.9f_r} = \varepsilon_1 - 0.186\beta/f_r \quad (4.51)$$

From (4.48),

$$\varepsilon_1 = \frac{0.8 \times 28.6f_r}{200\,000 \times 25.1\beta} = 4.56 \times 10^{-6}f_r/\beta \quad (4.52)$$

For unpropped construction, $\beta = 0.65$ (i.e., 35% redistribution) and $f_r = 156$ N/mm^2. From (4.52), $\varepsilon_1 = 1095 \times 10^{-6}$. From (4.51),

$$10^6\varepsilon_m = 1095 - 775 = 320$$

From (4.50),

$$w = 0.054 \text{ mm} \quad (4.53)$$

For propped construction, $\beta = 0.79$ and $f_r = 400$ N/mm^2. From (4.52),

$$\varepsilon_1 = 2310 \times 10^{-6}$$

From (4.51),

$$10^6\varepsilon_m = 2310 - 370 = 1940$$

From (4.50),

$$w = 0.33 \text{ mm} \quad (4.54)$$

These results show that the effect of tension stiffening is much greater (775 microstrain cf. 370) for unpropped construction, due to the lower strain in the top reinforcement; and the crack width then is low. The value $w = 0.33$ mm for propped construction might or might not be acceptable, depending on the environment and the floor finish.

It is of interest that the increase in the area of top longitudinal reinforcement from 12.06 cm^2 to 25.1 cm^2 is necessary in the propped beam to reduce deflections and in the unpropped beam to reduce crack width. It is true generally that for unpropped construction it is more important to check deflection than crack width, whereas for propped construction the converse is true.

4.6.2   Crack-width control in positive-moment regions

In the preceding worked example, the maximum midspan bending moment occurred with imposed load on one span only, when the moment at the internal support was not redistributed. The steel stresses f_{s4} were well below yield, for both propped and unpropped construction. In spans continuous at both ends, the working-load tensile stress is likely to be lower still. Crack widths at the lower surface of an encased beam usually require attention only in simply supported spans in which high-yield steel is used.

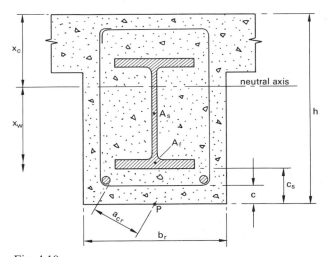

Fig. 4.18

The applicability of the formulae given in CP 110 (eqs. (4.43) and (4.44)) to a simply supported beam of the section shown in Fig. 4.18 is therefore considered. There will be at least two longitudinal bars near the bottom surface of the concrete encasement, so the widest crack is likely to occur at a point such as P in Fig. 4.18. For a given beam, bar size ϕ, and cover c, a trial distance a_{cr} is then known, and can be compared with a limiting value calculated from (4.43) in its rearranged form:

$$a_{cr} = \frac{h - x_c - 2c}{[3\varepsilon_m(h - x_c)/w] - 2} \tag{4.55}$$

As explained in Section 4.6, it is necessary to replace f_{yt} in (4.44) by $f_{s4}/0.58$, where f_{s4} is the maximum tensile stress in the steel member. According to CP 110, γ_m for concrete should be taken as 1.3 when analysing cross-sections for crack widths. The change from the usual value for serviceability analyses (1.0) reduces the design cube strength, and hence the value of E_c, and so causes a small increase in the modular ratio E_s/E_c. It can be shown that in composite beams the bottom-flange section modulus is not sensitive to modular ratio (e.g., in Table 4.3, doubling m reduces Z_{s4} by 6%), so that the effect on f_{s4} of this change in γ_m for concrete can be neglected. The value of b in (4.44) is that for the concrete rib (b_r in Fig. 4.18), and it is accurate enough to take A_{rt} as the area of the tension flange of the steel member, A_f. With these changes, (4.44) becomes

$$\varepsilon_m = \varepsilon_1 - \frac{0.7 b_r h x_w f_1}{A_f f_{s4}(h - x_c)} \times 10^{-3} \qquad (4.56)$$

where x is now written as x_c, as it is measured from the face of the member in compression.

An expression is now needed for ε_1, the strain at the level considered. For a point at depth x_w below the neutral axis, it is

$$\varepsilon_1 = \frac{x_w f_{s4}}{E_s(h - x_c - c_s)} \qquad (4.57)$$

where c_s is the bottom cover to the steel I-section.

4.6.3 Example. Crack-width control at midspan

We assume that the beam studied in the previous Example is encased in concrete such that $c_s = 50$ mm, and that 6-mm longitudinal bars are placed as shown in Fig. 4.18, with $c = 20$ mm. From Fig. 3.7, other dimensions are $h = 612$ mm, $b_r = 253$ mm, $A_f = 24.8$ cm^2. The pitch p of the two bars is $253 - 2 \times 23 = 207$ mm, and a_{cr} is found to be 103 mm.

In Section 3.7.2, the neutral-axis depth x_c for this beam was found to be 165 mm, so for point P, $x_w = 612 - 165 = 447$ mm. The maximum midspan stress f_{s4} (propped construction) was 235 N/mm^2. We will now check whether the two 6-mm bars provide sufficient crack control at point P, assuming the limiting crack width w to be

0·3 mm. From (4.57),

$$10^6\varepsilon_1 = 447 \times 235/0\cdot2 \times 397 = 1323$$

From (4.56),

$$10^6\varepsilon_m = 1323 - \frac{0\cdot7 \times 253 \times 612 \times 10^3}{2480 \times 235} = 1137$$

From (4.55),

$$a_{cr} = \frac{447 - 40}{(3 \times 0\cdot00114 \times 447/0\cdot3) - 2} = 131 \text{ mm}$$

This exceeds the value 103 mm found above, so the crack-width control is satisfactory.

4.7   Slender beams at the ultimate limit state

A procedure is required for checking the strength at the ultimate limit state of continuous beams to which the simple plastic theory is not applicable, due to the slenderness of the steel section. The following method is suggested.

The longitudinal moments due to the factored loads are found by elastic analysis. If it is assumed that the full composite section is effective, then compensation for the inaccuracy of the analysis should be made by redistributing the moments at the internal supports by about 20%, as discussed in Section 4.5.3. If time is available for a more accurate but complex analysis, it can be assumed that the concrete is cracked in negative-moment regions, and here the elastic properties of non-uniform beams given on p. 208 of Ref. 7 may be useful.

This method is likely to be used only when the steel section is slender, as defined in Section 4.4.2. It cannot then be assumed that the negative moment of resistance of the member can be developed without buckling. The stresses due to the calculated negative moment should be found by elastic theory for the cracked composite section, taking account of the method of construction of the beam. They should not exceed the design yield stress (f_{rd} or f_{sd}) in the slab reinforcement and the steel section, or such lower stress in the steel as may be specified in the 'steel' section of the revised BS 449.

In the midspan region it can be assumed that the slab prevents buckling of the top flange of the steel member, so that the calculated bending moment is satisfactory if it does not exceed the plastic moment of resistance of the composite member.

CHAPTER FIVE

Composite Columns and Frames

5.1 Introduction

The compressive strength per unit cost of concrete is higher than that of structural steel, so it would seem to be the more suitable material for use in columns. But when steel or composite beams are used in framed structures, it is usually necessary to provide steel stanchions, for the steelwork cannot be erected efficiently if the process has to be interrupted while columns of other materials are constructed. When a period of fire resistance is specified, the steel stanchions are usually encased in concrete. Until the 1950s, it was normal practice to use a wet mix of low strength and to neglect the contribution of the concrete to the strength of the column. Tests by Faber,[57] Stevens[58] and others then showed that savings could be made by using better-quality concrete and designing the column as a composite member. This led to the *cased-strut* method of design, described in Section 5.2. This empirical method was developed by stages from earlier design procedures for steel columns, and is not based on fundamental research on composite columns.

Progress since 1960 has been summarised by McDevitt and Viest.[59] The behaviour of isolated pin-ended composite columns subjected to axial loads and any combination of end moments about one or both axes is now well understood. Theoretical work, computer analysis, and a design method developed at Imperial College, London, are summarised in Section 5.3, and the method is applied to a concrete-encased steel member. It can also be used for concrete-filled steel tubes (Section 5.4), but at first sight has little in common with the design method for reinforced concrete columns given in CP 110.[9] Reinforced concrete frames are normally designed as if rigid jointed, and appropriate methods are included in both CP 110 and ACI 318. There is a long tradition of designing steel-framed structures for buildings as if the beams were simply supported at each column. The current British design methods for composite columns (Sections 5.2 and 5.3) are consistent with this approach. Research is still in progress on the best method of adapting the ultimate-strength method for use in a composite frame in which the beam–column joints cannot be assumed to act as pins. This is the subject of Sections 5.5 and 5.6.

Progress in the USA has taken a different route. The tendency has been to consider the steel member, whatever its shape, as reinforcement to the concrete. The current design method for reinforced concrete columns, ACI 318[60] is applicable also to encased stanchions and to concrete-filled tubes.

The methods are illustrated by an example, in which a column length is designed to support five storeys of the building structure used to illustrate Chapters 3 and 4.

5.2   Cased-strut design to BS 449: Part 2[10]

The method considers only a particular type of encasement, suitable for protecting a steel stanchion of I section against fire, and so is not applicable to concrete-filled tubes, nor to members where the concrete section has heavy longitudinal reinforcement or is substantially larger than the steel section.

The concrete encasement must have a minimum 28-day cube strength of 21 N/mm^2, and must provide a minimum cover of 50 mm to the steel member. It must be reinforced with wire binders at not more than 150 mm spacing. No account is taken of the longitudinal bars that are usually provided to support the binders.

The allowable stresses given in BS 449 must not be exceeded at working load which, in limit-state terminology, is equivalent to the design loading for the serviceability limit state. For an uncased steel strut of effective length L and cross-sectional area A_s, the condition to be satisfied at all cross-sections is

$$\frac{N}{N_a} + \frac{f_f}{f_{af}} \not> 1 \qquad (5.1)$$

where N is the axial load on the member, N_a is the allowable axial load, given by

$$N_a = A_s f_{ac} \qquad (5.2)$$

f_f is the resultant compressive stress due to bending about both the x- and y-axes, and f_{af} is the appropriate allowable compressive stress for members in bending.

The allowable axial stress f_{ac} is determined from the yield stress of the steel and the greater of the two slenderness ratios of the member, L_x/r_x and L_y/r_y. The stress f_{af} depends on the yield stress, the minor-axis slenderness ratio, and the ratio of the depth of the member to the mean flange thickness, and so includes allowance for lateral–torsional buckling.

The presence of concrete encasement is allowed for in two ways. The least radius of gyration is increased to $0.2(b + 100)$ mm, where b is the breadth of a flange of the steel member, which gives a corresponding increase in f_{ac}, but does not alter f_{af}. The allowable axial load is increased by assuming that the gross area of concrete, excluding cover in excess of 75 mm, may resist a compressive stress $f_{ac}/0.19f_{ab}$, where f_{ab} is the allowable compressive stress for *short* members in bending (so that it differs from f_{af} in not including a reduction for lateral–torsional buckling). The coefficient 0.19 was probably derived from the contemporary allowable stress for Grade 21 concrete in compression, 760 lb/in², which is $(1/0.19)$ N/mm². The factor f_{ac}/f_{ab} scales down this stress by the same amount that the allowable axial stress in the steel is reduced to allow for the slenderness of the member. So (5.2) becomes

$$N_a = A_s f_{ac} + A_g f_{ac}/0.19 f_{ab} \qquad (5.3)$$

where A_g is the gross area of concrete, as defined above. There is the further restriction that the axial load on the member may not exceed twice that which would be permitted on the uncased strut. A worked example is given in Section 5.2.1.

Tests on cased struts under axial and eccentric load show that this method gives a very uneven and usually excessive margin of safety. For example, Jones and Rizk[61] quote load factors ranging from 4·7 to 6·7, and work by Stevens[62] and Faber[57] supports this conclusion. There are five main reasons why the method is unsatisfactory, even for the idealised 'pin-ended' column which so rarely occurs in practice.

(1) Encased steel columns behave in tests like reinforced concrete columns, for which it is well established that the modular ratio m is irrelevant at ultimate load. Equation (5.3) above is equivalent to assuming that $m = 30$ for Grade 43 steel and $m = 44$ for Grade 50.

(2) The allowable axial load should be related to a slenderness calculated from the least lateral dimension of the concrete cross-section, or from the transformed cross-section, not from the breadth of the steel flange.

(3) Tests confirm what has long been recognised for reinforced concrete, that the axial compressive stress need not be reduced below the value for very short columns until L/r exceeds about 50; but in the cased-strut method, reductions begin when $L/r > 0$.

(4) The bending moment at mid-length of a column is influenced both by the L/r ratio and by the interaction between the axial load and the lateral deflection due to end moments. Equation (5.1) allows for the first effect, but not for the second.

(5) The concrete encasement prevents lateral–torsional buckling of a column length, so that it is over-conservative to relate the bending stress f_f in the steel to an allowable stress f_{af} that may be well below the appropriate stress, which is f_{ab}. For example, if the column section shown in Fig. 5.1 were used in a column of effective length 8 m, the ratio f_{af}/f_{ab} given by BS 449 for Grade 50 steel would be 0·73.

For all its faults, this design method is widely used, mainly because it is much simpler than the more rational methods now available, as the following descriptions of them will show.

5.2.1   Example (continued). Design by the cased-strut method

The two-bay building structure described in Section 4.2 is assumed to consist of six or more identical storeys, braced against sidesway by walls at the ends of the building. A design is required for an external column length to support five storeys of the floor structure already designed. The beams (Fig. 4.1) are attached to the column by simple cleated connections and it is assumed (for simplicity) that there is no additional load from the external walls.

The strengths of the materials are as used in Chapter 4:

structural steel, Grade 50: $f_y = 350$ N/mm², $\gamma_m = 1\cdot15$, so $f_{sd} = 304$ N/mm²;

reinforcement: $f_{rd} = 410/1\cdot15 = 356$ N/mm²;

concrete: $f_{cu} = 30$ N/mm² at 28 days, $\gamma_m = 1\cdot5$.

The γ-values $1\cdot15$ and $1\cdot5$ are for the ultimate limit state.

Column 'design' methods are in fact methods of determining whether a proposed member is strong enough. To enable comparisons to be made between methods, the same member will be checked by several methods, even though it is not the most economical design in every case. Its cross-section is shown in Fig. 5.1. The reinforcement is neglected in design. Properties of the steel cross-section are

$A_s = 66\cdot4$ cm²,

$I_{sx} = 5263$ cm⁴, $r_x = 8\cdot89$ cm, $Z_x = 510$ cm³,

$I_{sy} = 1770$ cm⁴, $r_y = 5\cdot16$ cm, $Z_y = 174$ cm³.

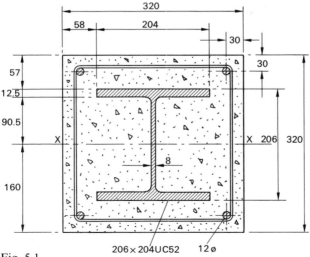

Fig. 5.1

For use in (5.3), the area A_g may be taken as 32^2, or 1024 cm^2; the net concrete area is $A_c = 1024 - 66 = 958$ cm^2. The weight of the column, g_c, is assumed to be 3 kN/m.

The vertical shear at the end of the beam (Fig. 4.1) is greatest when the negative moment at the internal support is least, so its precise value depends on the extent of cracking at that support and, if un-propped construction is used, on the sequence of construction. Any error in the vertical shear due to use of the wrong value of the negative moment is certain to be much less than the degree of approximation implied by the use of the nominal eccentricities of loading given in BS 449. Except at the lower levels of a tall structure, where column design is governed by the sum of the vertical shears in the beams, any reasonable approximation of the value of the negative moment will be accurate enough.

We now assume that propped construction is used. When imposed load acts on one span only, the negative moment (Section 4.5.4) is 414 kNm, which is reduced to 327 kNm after redistribution of 21%. The total load on the beam is 45 kN/m, so the load on the external column is

$$V = \tfrac{1}{2} \times 10\cdot5 \times 45 - 327/10\cdot5 = 236 - 31 = 205 \text{ kN} \quad (5.4)$$

According to BS 449, this load should be assumed to act at an eccentricity of 100 mm from the face of the steel member. The moment applied to the column is assumed to be shared equally between the upper and lower column lengths, so the design moment for one length is given by

$$M_x = \tfrac{1}{2} \times 205(0\cdot103 + 0\cdot1) = 20\cdot8 \text{ kNm} \quad (5.5)$$

The storey height is 4 m, so the unfactored weight of a column supporting five storeys is $5 \times 4 \times 3 = 60$ kN, and the total axial load is

$$N = 5 \times 205 + 60 = 1085 \text{ kN} \quad (5.6)$$

The least radius of gyration of the cased section is

$$0\cdot2(20\cdot4 + 10) = 6\cdot08 \text{ cm}$$

The beam is connected to the column flange, so that buckling about the weaker axis would occur in the plane of the external wall of the building. In this plane, the ends of the column are assumed to be held in position but not restrained in direction, so the effective length should be taken as the actual floor-to-floor height (4 m) and the slenderness is

$$L/r_y = 400/6 \cdot 08 = 66 \tag{5.7}$$

For this value and the steel member used here, BS 449 gives $f_{ac} = 158 \text{ N/mm}^2$ and $f_{af} = f_{ab} = 230 \text{ N/mm}^2$ for Grade 50 steel. So from (5.3),

$$N_a = 158(66 \cdot 4 + 1024/0 \cdot 19 \times 230) \div 10$$
$$= 1049 + 370 = 1419 \text{ kN} \tag{5.8}$$

The maximum bending stress is

$$f_f = M_x/Z_x = 20\ 800/510 = 40 \cdot 8 \text{ N/mm}^2$$

Substituting in (5.1),

$$1085/1419 + 40 \cdot 8/230 = 0 \cdot 765 + 0 \cdot 177 = 0 \cdot 94$$

This is less than $1 \cdot 0$, so the strength of the column is sufficient, but not excessive.

5.3   Ultimate-strength design method of Basu and Sommerville

As originally presented, the method is applicable to pin-ended uniform composite members, having a cross-section symmetrical about two perpendicular axes, and subjected to axial load plus any combination of end moments M_x and M_y, which need not be the same at the two ends of the member. The principal assumptions and limitations of the method are discussed here, and all equations needed for the subsequent worked examples are given; but the reader seeking a detailed account is referred to the original paper.[63]

Loading. The column length is assumed to be loaded in such a way that plane sections remain plane, without bond failure near the ends, and the maximum load is assumed to be reached without local or lateral–torsional instability. The relevant material properties are the design yield stresses of the steel and the reinforcement, f_{sd} and f_{rd}, and the design cube strength of the concrete, $f_{cu}/1 \cdot 5$.

Short-term loading is considered first. It is defined by the axial load N, and the end moments M_x, $\beta_x M_x$, M_y and $\beta_y M_y$, so allocated that for both axes, $-1 \leqslant \beta \leqslant 1$, with positive β corresponding to single-curvature bending (Fig. 5.2). The effects of transverse shear forces are found to be negligible in tests, and are not considered in the design method. Biaxial bending is considered only near the end of the calculation, after the failure loads N_x (due to N and the x-moments) and N_y (due to N and the y-moments) have been calculated. The method of calculation is the same for both axes, so symbols such as M and β will now be used without suffixes x or y, and should be assumed to have the values appropriate to the axis under consideration.

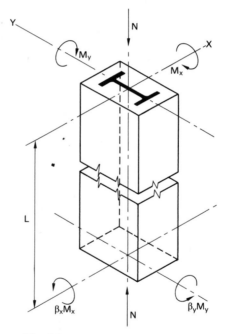

Fig. 5.2

5.3.1   Properties of the cross-section, and the interaction curve

For each axis of bending, the properties of the cross-section of the column are represented by four parameters: N_u, r, α, and M_p. These are now discussed in turn. The first is the ultimate axial load for a short column (the squash load), given by

$$N_u = A_s f_s + A_r f_r + A_c f_u \tag{5.9}$$

where A_c is the actual area of concrete and f_u is the strength of concrete in axial compression in a column, assumed to be

$$f_u = 0.83 f_{cu}/\gamma_m \tag{5.10}$$

For a slender all-steel column, the ratio K_1 of the ultimate axial load N_a to the squash load depends mainly on the slenderness ratio L/r. For elastic (Euler) buckling, it is

$$K_1 = N_a/N_u = \pi^2 E_s/f_y (L/r)^2 \tag{5.11}$$

Due to initial curvature and local yielding of the member, the curve for a real steel column is lower, as sketched in Fig. 5.3. One such curve, due to Perry and Robertson, is used in BS 449.

For a composite cross-section, it was found possible to define an equivalent radius of gyration, r, such that the ratio of its elastic critical load to its squash load is given by (5.11). The expression includes Young's modulus for concrete, E_c. Curves relating K_1 to L/r were computed for composite columns, taking account of inelastic behaviour and initial curvature, and were found to have least scatter when E_c was assumed to be $300 f_{cu}/\gamma_m$. For $f_{cu} = 30$ N/mm^2, this implies that $E_c = 6.0$ kN/mm^2, whereas the common use of $m = 7.5$ for short-term loading gives $E_c = 26.7$ kN/mm^2. Basu and Sommerville did not comment on this discrepancy. Their value is influenced by assumptions made in their computations, and is in effect an average tangent modulus, that takes account of the inelastic behaviour of concrete as failure is approached. It is therefore much lower than the secant modulus for short-term loading. The assumption $E_c = 300 f_{cu}/\gamma_m$ led to this expression for r:

$$r^2 = (f_y^* I_s + 0.24 f_{cu} I_c)/N_u \tag{5.12}$$

where I_s and I_c are the second moments of area of the steel and the concrete in the cross-section and f_y^* is the yield stress for mild steel, here taken as 250 N/mm^2.

The curve finally taken for K_1 was a lower bound of the computed results calculated using (5.12) for r. It is close to the Perry–Robertson curve except at L/r ratios exceeding about 150 (Fig. 5.3). Here the

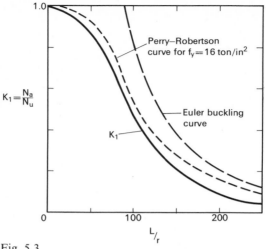

Fig. 5.3

discrepancy is such that calculations for certain steel sections lead to the improbable conclusion that the ultimate axial load is reduced when a concrete casing is added. There is at present not enough experimental data on such slender columns to justify raising the curve in this region. The inconsistency is not important, for composite columns with L/r exceeding 150 are rarely used.

The first objective of the design calculations is to obtain an inter-action curve for the ultimate strength of the member under eccentric loading. The curve relates N/N_u to M/M_p, where M is the maximum end moment and M_p is the plastic moment of resistance in absence of axial load. For steel members the curves may be assumed to depend only on L/r and β. They can be idealised as a straight line and a parabolic arc, as shown in Fig. 5.4, defined by three parameters K_1, K_2, and K_3. The addition of concrete to a steel member has the effect of altering the shape of the curve, and the amount of concrete has to be reflected in the calculation of K_2 and K_3. This can be done by defining a *concrete contribution factor*, α, by

$$\alpha = A_c f_u / N_u \tag{5.13}$$

Comparison with (5.9) shows that α is simply the fraction of the squash load of the member contributed by the concrete.

Plastic moment of resistance.   The fourth property of the cross-section is the plastic moment of resistance at zero axial load, M_p. Tests show that under a small axial load the flexural strength of a composite member can exceed M_p (dashed line in Fig. 5.4), in the same way that axial prestress strengthens a reinforced concrete beam. It is not practicable to take account of this in column design, so the line $M/M_p = 1$ forms part of the interaction curve, and the maximum axial load at which M_p can be reached is given by the parameter K_2 (Fig. 5.4), which is discussed later.

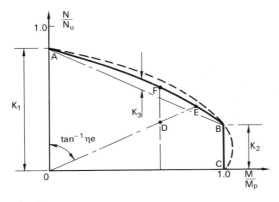

Fig. 5.4

The calculation of M_p is based on rectangular stress blocks, exactly as for composite beams. The same assumptions (Section 3.4) are made, except that Basu and Sommerville take the mean compressive stress in the concrete as $0.67f_{cu}/\gamma_m$, or $0.44f_{cu}$, whereas $0.40f_{cu}$ is used in Chapter 3, in accordance with CP 110 for reinforced concrete.

In taking account of the longitudinal reinforcement, the following assumptions can usually be made, so avoiding tedious trial and error in finding the position of the neutral axis:

(1) The amount of reinforcement is small, and the areas in tension and in compression are equal.
(2) The design yield stress of the reinforcement in compression is assumed to be the same as in tension.
(3) The area of concrete in compression is not reduced to allow for the area occupied by steel in compression.

The first assumption is usually correct for composite columns. The second differs from CP 110, in which the design compressive stress in column reinforcement is reduced slightly to improve the margin of safety against buckling of the bars, and the third assumption is slightly on the unsafe side. In their study of the accuracy of the method, Basu and Sommerville found errors in M_p of up to 4% on the unsafe side.

The assumptions are a useful simplification, and are adopted here, but to provide some compensation for them, and for uniformity with composite beams and reinforced concrete columns, it is suggested that the mean compressive stress in concrete should be taken as $0.40f_{cu}$, not $0.44f_{cu}$, when calculating M_p.

For major-axis bending of an encased I-section, the neutral axis may be above or within the compression flange, or in the web. In practice, it is almost always in the web, as shown in Fig. 5.5. As in Fig. 3.6(c), it is convenient to split up the stress block for the I-section, and here A_f is the area of one flange, and $d_{wc}t_w$ the area of web in compression.

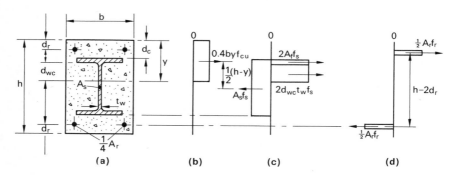

Fig. 5.5

When the three assumptions given above are made, the depth of the neutral axis, y, is most simply obtained by noting that the compressive force in the concrete is equal to the tensile force in an area $(h - 2y)t_w$ of web, since other longitudinal forces cancel out. Thus

$$0.4byf_{cu} = (h - 2y)t_w f_s$$

whence

$$y = \frac{ht_w}{2t_w + 0.4b(f_{cu}/f_s)}$$

provided that

$$y \geqslant d_c + t_f$$

(5.14)

Taking moments about the line of action of the force in the concrete leads to

$$M_p = [\tfrac{1}{2}A_s(h - y) - A_f(d_c - d_{wc}) - d_{wc}t_w(y - d_{wc})]f_s \\ + A_r f_r(\tfrac{1}{2}h - d_r) \quad (5.15)$$

For minor-axis bending of an encased I-section, the neutral axis almost always intersects the flanges, as shown in Fig. 5.6(a). Resolving longitudinally,

$$0.4hyf_{cu} = A_s f_s - 4(y - d_c)t_f f_s$$

whence

$$y = \frac{A_s + 4t_f d_c}{4t_f + 0.4h(f_{cu}/f_s)}$$

(5.16)

Taking moments as before gives

$$M_p = [\tfrac{1}{2}A_s(b - y) - 2(y - d_c)t_f d_c]f_s + A_r f_r(\tfrac{1}{2}b - d_r) \quad (5.17)$$

Results similar to (5.15) and (5.17) can be obtained[63] in the same way for other positions of the neutral axis and for concrete-filled steel tubes (p. 174).

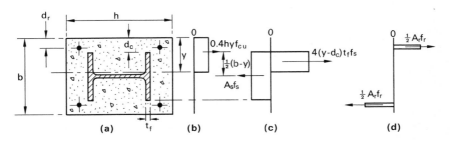

Fig. 5.6

Expressions for K_2 and K_3.   Interaction curves similar to the dashed line in Fig. 5.4 were computed for about one hundred composite columns, and the corresponding values of K_2 and K_3 were obtained. These were found to depend on L/r, α, β, and the type of cross-section considered. The following empirical expressions for K_2 and K_3 were then developed,[63] giving design values on the safe side of those computed.

For encased sections and filled tubes,

$$\frac{K_2}{K_{20}} = \frac{90 - 25(2\beta - 1)(1\cdot8 - \alpha) - L/r}{30(2\cdot5 - \beta)} \qquad (\geqslant 0 \text{ and } \leqslant 1) \tag{5.18}$$

where K_{20}, the value when $L/r = 0$, is given by

$$K_{20} = 0\cdot9\alpha^2 + 0\cdot2 \qquad (\leqslant 0\cdot75) \tag{5.19}$$

For filled rectangular tubes and for major-axis bending of encased sections, it was found that

$$K_3 \simeq 0 \tag{5.20}$$

For minor-axis bending of encased sections,

$$K_3 = 0\cdot425 - 0\cdot075\beta - 0\cdot005L/r \tag{5.21}$$

between the limits

$$\left. \begin{array}{l} K_3 = 0\cdot2 - 0\cdot25\alpha \qquad (\text{but } \not< 0) \\ K_3 = -0\cdot03(1 + \beta) \end{array} \right\} \tag{5.22}$$

Use of interaction diagram.   In the preceding expressions for K_2 and K_3, α and r are properties of the cross-section, L is a dimension of the structure, and β is a function of the loading, so that the interaction curve has to be constructed afresh for each new trial cross-section, and each new type of loading.

In practice, a designer needs to know whether a given member can resist axial load N and end moments M and βM. One method would be to construct the failure envelope (ABC on Fig. 5.4) and plot the point (M/M_p, N/N_u). If this lies within the envelope (e.g., point D), the member is strong enough. The algebraic conditions for this are

$$M \leqslant M_p$$
$$N/N_u \leqslant K_1 - (K_1 - K_2 - 4K_3)(M/M_p) - 4K_3(M/M_p)^2 \tag{5.23}$$

If the ultimate strength of the member is required, it may be assumed that the ratio of M to N remains constant as the load increases, which is correct for pin-ended columns but not for columns in rigid-jointed frames. Let the ratio be

$$e = M/N \tag{5.24}$$

and let

$$\eta = N_u/M_p, \tag{5.25}$$

so that $\tan \angle AOD = \eta e$, and the line ODE on Fig. 5.4 represents increasing load, with failure at E. Then N/N_u can be calculated as follows.

If $\eta e \geqslant 1/K_2$, then

$$N/N_u = 1/(\eta e) \tag{5.26}$$

If $\eta e < 1/K_2$ and $K_3 \simeq 0$,

$$N/N_u = K_1/[1 + (K_1 - K_2)\eta e] \tag{5.27}$$

If $\eta e < 1/K_2$ and $K_3 \neq 0$,

$$\left. N/N_u = \frac{-K_4 + [K_4^2 + 16K_1 K_3 (\eta e)^2]^{1/2}}{8K_3(\eta e)^2} \right\}$$

where

$$K_4 = 1 + (K_1 - K_2 - 4K_3)\eta e \qquad (5.28)$$

An alternative assumption, which leads to simpler formulae, is that the maximum end moment M remains constant as the axial load increases. The ratio N/N_u is then given by

$$N/N_u = K_1 - (K_1 - K_2 - 4K_3)(M/M_p) - 4K_3(M/M_p)^2 \tag{5.29}$$

5.3.2 Biaxial bending

A method for computing the strength of a composite column under a combination of axial load and biaxial moments has been developed by Virdi and Dowling,[64] and compared with the results of tests on nine encased columns. Allowance was made in the theory for residual stresses and initial lack of straightness of the column by assuming an initial bow at midlength of $6 \times 10^{-5}(L^2/D)$ about both axes, where D is the overall depth of the steel section in the plane considered. (This is incorrectly given in Ref. 64 as $6 \times 10^{-4}L^2/D$.) There was broad agreement between theory and tests, with some discrepancies that were probably due to the test columns being less imperfect than was assumed.

The theory is too complex for use in design, so the tests were also compared with the predictions of the formula

$$\frac{1}{N_{xy}} = \frac{1}{N_x} + \frac{1}{N_y} - \frac{1}{N_{ax}} \tag{5.30}$$

in which N_{xy} is the biaxial failure load, N_x and N_y are the failure loads in uniaxial bending given by the method described above, and N_{ax} is the failure load under axial loading when bending is constrained to occur about the x-axis (i.e., $N_{ax} = K_{1x}N_u$). This formula was found to be uniformly conservative, in one case by almost 50%. The errors arise partly in (5.30) above, which is a rough approximation to the true interaction surface, and partly in N_x, N_y and N_{ax}, which are themselves conservative due to the cumulative effect of assumptions made in deriving the design method.

For short columns, the errors are small enough to be acceptable, but a more economical method for long columns is needed.

Initial imperfections. These may exist in either plane of bending, and were allowed for in the derivation of K_1, K_2, and K_3 by assuming an initial bow at mid-length of $6 \times 10^{-5}L^2/D$, as explained above. When the applied moments act only about the weaker axis (the y-axis), then $N_x = N_{ax}$, and (5.30) gives $N_{xy} = N_y$, showing that the effect of the x-axis imperfections is neglected. But when the applied moments act only about the x-axis, N_x is less than N_{ax}, and (5.30) gives a biaxial failure load N_{xy} which may be less than N_x, due to the y-axis imperfections.

Modified interaction formula. The interaction formula given in the draft Bridge Code[15] looks identical with (5.30) but in fact is less conservative, due to a redefinition of N_x and N_y, proposed by Virdi and Dowling. This change is now explained, with reference to a column at failure under axial load N and greater end moments M_x

and M_y. If N is to satisfy the biaxial criterion (5.30), the point $(N/N_u, M_x/M_{px})$ must lie within the uniaxial failure envelope for the x-plane (e.g., point D in Fig. 5.4). In the method outlined above, N_x is calculated at constant eccentricity (e_x), and so is given by point E. In the revised method, N_x is defined as the x-axis failure load at constant bending moment (M_x). This gives point F in Fig. 5.4, which corresponds to a higher value of N_x. The definition of N_y is changed in the same way. These values of N_x and N_y are found from (5.29), rather than from (5.26) to (5.28), and (5.30) then gives a higher value of N_{xy} than before.

5.3.3   Long-term loading

The stress–strain curve for concrete used in the computations on which the design method is based took no account of creep under sustained load. This effect was simulated by repeating the calculations with the stiffness of the concrete halved, which corresponds to doubling the modular ratio. The ratio of the short-term strength to the long-term strength was thus found. It was recommended[64] that the long-term load and moments applied to the column should be multiplied by this ratio to compensate for the effects of creep of concrete. The calculations led to the following values of the magnification factor, m:

for filled tubes,

$$m = 1 + 3\alpha/8 \qquad (5.31)$$

for encased sections,

$$m = 1 + L/250r \qquad (5.32)$$

with the restriction

$$m = \not> 1 + [0.11/(\lambda + 0.13)] \qquad (5.33)$$

where $\lambda = f_y^* I_s/f_{cu} I_c$ and $f_y^* = 250 \text{ N/mm}^2$.

There is little experimental data to support this procedure, and there is as yet no agreed method of deciding what proportion of the load should be assumed to be 'long term'. The normal distinction between dead and imposed load treats as 'imposed' certain types of loading (e.g., office equipment) which may be present for years on end. It would be consistent with the assumptions on which the method is based to treat as long-term load the whole dead load and a proportion of the imposed load, particularly in buildings such as warehouses.

5.3.4   Simplication of the Basu and Sommerville method

In Ref. 63, graphs are given for K_1, K_{20}, K_2/K_{20}, and K_3; but even when these are used the time required to design a column length is longer than when the cased-strut method is used.

Another reason why it may at first be unattractive to designers is that there are many differences between it and other column-design methods with which they may be familiar. For example, in allowing for column slenderness, it modifies the shape of the interaction diagram, whereas both CP 110 and the Joint Committee method[48] use the concept of additional moments. This difference is fundamental; but there are others where changes could be made to simplify the method and bring it more into line with CP 110. These are now discussed.

The ultimate compressive strength of concrete in a cross-section under pure axial load, f_u, is taken as $0.83 f_{cu}/\gamma_m$, which is $0.55 f_{cu}$ when γ_m is 1.5. The value in CP 110 is $0.45 f_{cu}$. This is based on extensive studies of all available data by the European Committee for Concrete and the Cement and Concrete Association.

Basu studied this subject by using the program on which the design method is based to compute failure loads for forty columns for which test results were available, taking f_u as $0.45 f_{cu}$ and $0.55 f_{cu}$ in turn. Of the short columns, only five were axially loaded, and for these the mean ratios of test load to computed load were 1.04 and 0.97 respectively for the two values of f_u. The higher value was adopted for the design method because it gave better overall correlation between test and computed results.

For sustained loading, CP 110 states only that 'appropriate allowance for shrinkage and creep should be made'. There is nothing corresponding to the magnification factors of Section 5.3.3. Designers are likely to continue the current practice of making no special allowance for creep when designing reinforced concrete columns.

In the Basu and Sommerville method for filled tubes, the extra load due to the magnification factor is roughly equivalent to the extra axial strength due to the higher value of f_u. If the long-term load is half the total load, the increase in axial load is, from (5.31), $3\alpha/16$, or $19\alpha\%$. The increase in axial strength due to the higher value of f_u is $(0\cdot10/0\cdot45)\alpha$, which is $22\alpha\%$. This suggests that the magnification factor could be omitted if f_u is taken as $0\cdot45f_{cu}$, with the further advantage that no separate calculation of the 'long-term' loading is needed.

Comparisons with test data are needed to establish whether this assumption is safe for encased sections, in which the effect of creep on buckling load is greater than in filled tubes. It can have a significant effect on design, particularly in members with large end moments and a high proportion of long-term load, as shown in the example that follows.

It is common in practice for a column length to be subjected to end moments about the major axis only. The Basu and Sommerville method for such columns involves calculations for bending about both axes and the use of the interaction formula (5.30), which is known to be conservative. Even so, N_{xy} is often found to be only a little below N_x.

The importance of the biaxial-bending calculation increases as the ratio of L_y/r_y to L_x/r_x increases, and it is certainly necessary when the ratio exceeds two or three. But for the columns most likely to be used in practice, the ratio is less than $1\cdot5$, and it should be possible to show that for such columns loaded about the major axis only, there is no need to allow for minor-axis bending.

Finally, design of the great majority of encased columns could be done from tables similar to those available for filled tubes.[65] Only 32 cross-sections need be considered: the standard Universal Column sections with rectangular encasement giving 50 mm of cover and four longitudinal 12-mm reinforcing bars with $f_r = 410$ N/mm^2. Values of ultimate axial load N_x and N_y could be tabulated for appropriate values of effective length L and ratio e_x/h or e_y/b, where e is the greater end eccentricity. Tables for $f_y = 250$ and 350 N/mm^2, $f_{cu} = 25, 30$, and 40 N/mm^2, and $\beta = +1, +0\cdot5, 0$, and $-0\cdot5$, should be sufficient. Their total length would be about the same as that of the tables for filled tubes.

5.3.5 Eccentricity of loading in 'simple' design

When a steel or composite beam is attached to the steel member of a composite column by a simple connection not intended to resist bending moment (such as a bolted bottom bracket and web cleat), some assumption must be made about the eccentricity at which the load is applied to the column. The existing rule in BS 449 (at the centre of the bearing or 100 mm from the face of the section, whichever is greater) was intended for steel columns. It should be equally applicable to concrete-filled tubes. It may be unsafe for encased columns, particularly if the cover to the steel member (d_c in Fig. 5.5) is large or if the composite beam is encased, because then the connection is stiffened by the surrounding concrete.

Some allowance for this can be made by interpreting the phrase 'the face of the section' in BS 449 to mean 'the face of the *composite* section' when the column is designed as a composite member. This does not alter the eccentricity for filled tubes, and is being proposed in the revision of BS 449 now in preparation. It is further discussed in Section 5.5.1.

5.3.6 Example (continued). Design by ultimate-strength methods

The method of Basu and Sommerville will now be used to find out whether the column section shown in Fig. 5.1 is suitable to support five storeys of the floor structure already designed. All data is as in Section 5.2.1, except that the loading is that for the ultimate limit state.

Loads. On p. 112 it was shown that at flexural failure the bending-moment diagram for the beam of the present example lies between the limits shown in Fig. 4.7, and when designing the beam for vertical shear at E, the upper curve was used. Similarly, for maximum load at the external support G, the lower curve is appropriate. The maximum positive moment of 770 kNm corresponds to a simply supported span of $(8 \times 770/70)^{1/2}$, or 9·38 m, and hence to a vertical shear at G of $\frac{1}{2} \times 9\cdot38 \times 70 = 328$ kN, of which 103 kN is due to dead load.

From Section 5.3.4 and Fig. 5.1, the eccentricity is $160 + 100 =$

260 mm. Assuming that the applied moment is resisted equally by the upper and lower column lengths, the design moments for one column length are

$$\left.\begin{array}{lll} \text{dead load} & M_g = \frac{1}{2} \times 103 \times 0.26 = 13.4 \text{ kNm} \\ \text{imposed load} & M_q = \frac{1}{2} \times 225 \times 0.26 = 29.3 \text{ kNm} \end{array}\right\} \text{(5.34)}$$

The factored weight of a column supporting five storeys is $5 \times 4 \times 3 \times 1.4 = 84$ kN, so the axial loads are

$$\left.\begin{array}{lll} \text{dead load} & N_g = 5 \times 103 + 84 = 600 \text{ kN} \\ \text{imposed load} & N_q = 5 \times 225 \quad\quad = 1126 \text{ kN} \end{array}\right\} \text{(5.35)}$$

If both the upper and lower beams are fully loaded (Fig. 5.7), the column AC is bent in symmetrical double curvature ($\beta = -1$). Its ultimate strength is reduced slightly if M_2 is given its *minimum* possible value, which is 13·4 kNm. If AB is fully loaded, $M_1 = 13.4 + 29.3 = 42.7$ kNm, so that

$$\beta = -13.4/42.7 = -0.314 \quad\quad (5.36)$$

The beams are assumed to provide no rotational restraint to the column, so its effective length is its actual length, 4·0 m.

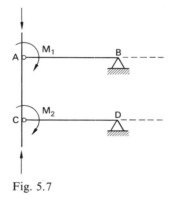

Fig. 5.7

Properties of the cross-section. To illustrate the method, the longitudinal reinforcement is now assumed to consist of four 12-mm bars of steel with $f_{rd} = 356$ N/mm^2, so that $A_r = 4.5$ cm^2. The net area of concrete is

$$A_c = 1024 - 66.4 - 4.5 = 953 \text{ cm}^2 \quad\quad (5.37)$$

From (5.9), (5.10), and data given in Section 5.2.1,

$$f_u = 0.83 \times 30/1.5 = 16.6 \text{ N/mm}^2 \tag{5.38}$$

and

$$N_u = 6.64 \times 304 + 0.45 \times 356 + 95.3 \times 16.6$$
$$= 2019 + 161 + 1582 = 3762 \text{ kN} \tag{5.39}$$

The concrete contribution factor is

$$\alpha = 1582/3762 = 0.42 \tag{5.40}$$

From Fig. 5.1, the second moment of area of the reinforcing bars about each axis is

$$4.5 \times (16 - 3)^2 = 764 \text{ cm}^4$$

so for the total steel section,

$$\left. \begin{array}{l} I_{sx} = 5263 + 764 = 6027 \text{ cm}^4 \\ I_{sy} = 1770 + 764 = 2534 \text{ cm}^4 \end{array} \right\} \tag{5.41}$$

For the whole cross-section, $I_x = I_y = 32^4/12 = 87\,380 \text{ cm}^4$, so from (5.41) the second moments of area of the concrete cross-section are

$$I_{cx} = 81\,350 \text{ cm}^4 \qquad I_{cy} = 84\,850 \text{ cm}^4 \tag{5.42}$$

Substituting f_y^*, f_{cu} and N_u into (5.12),

$$r^2 = 10^{-6}(66.4 I_s + 1.914 I_c) \text{ mm}^{-2} \tag{5.43}$$

so from (5.41) and (5.42)

$$r_x = 74.6 \text{ mm} \qquad r_y = 57.5 \text{ mm} \tag{5.44}$$

To calculate the plastic moment of resistance about the x-axis, the relevant quantities (Figs. 5.1 and 5.5) are $b = h = 320$ mm, $t_w = 8$ mm, $d_c = 57$ mm, $t_f = 12.5$ mm, $d_r = 30$ mm, $f_{sd} = 304$ N/mm^2, $f_{rd} = 356$ N/mm^2, $f_{cu} = 30$ N/mm^2, $A_f = 25.5$ cm^2.

From (5.14),

$$y = 89.4 \text{ mm} \tag{5.45}$$

so $d_{wc} = 89.4 - 57 - 12.5 = 20$ mm, and the use of (5.14) is correct, since $y > d_c + t_f$. Substituting in (5.15),

$$M_{px} = 222 \text{ kNm} \tag{5.46}$$

The contribution of the longitudinal reinforcement to M_{px} is 21 kNm, or about one-tenth. Increasing the bar diameter from 12 to, say, 25 mm would increase this contribution to about 80 kNm, and the total M_{px} to about 270 kNm. Thus composite sections, like reinforced concrete, have the practical advantage over steelwork that substantial adjustments to the strength of the member can be made at a late stage in the design process, after the steel frame and the concrete outlines have been 'frozen'.

The interaction diagram.   With $L = 4$ m and r from (5.44),

$$L/r_x = 53\!\cdot\!6 \qquad L/r_y = 69\!\cdot\!6 \tag{5.47}$$

From Fig. 5.3,

$$K_{1x} = 0\!\cdot\!84 \qquad K_{1y} = 0\!\cdot\!75 \tag{5.48}$$

Now $\alpha = 0\!\cdot\!42$, so from (5.19),

$$K_{20} = 0\!\cdot\!359 \tag{5.49}$$

There are no applied moments about the y-axis, so K_2 is needed for the x-axis only. From (5.18) with $\beta = -0\!\cdot\!314$, $K_{2x}/K_{20} = 1\!\cdot\!1$ (but $\leqslant 1$) so $K_{2x}/K_{20} = 1\!\cdot\!0$.
So from (5.49),

$$K_{2x} = 0\!\cdot\!359 \tag{5.50}$$

The interaction diagram for x-axis bending is as shown in Fig. 5.8, since $K_3 = 0$ from (5.20).

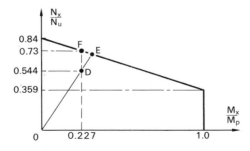

Fig. 5.8

Long-term loading. The magnification factors are given by (5.32) and (5.47) as

$$m_x = 1 \cdot 21 \qquad m_y = 1 \cdot 28$$

The limiting values are given by (5.33) as

$$m_x = 1 \cdot 15 \qquad m_y = 1 \cdot 29$$

so the values adopted are

$$m_x = 1 \cdot 15 \qquad m_y = 1 \cdot 28 \tag{5.51}$$

Basu and Sommerville give no guidance on how these values should be used. It seems reasonable to use the greater value, m_y, for the axial load, and to use m_x for the x-axis moments, since the magnification of these due to creep depends on L/r_x, not on L/r_y. But it is simpler in practice to use the greater value for all forces and moments due to dead load, and this is done here.

Here it is assumed that half of the imposed load· is long term. From (5.34), (5.35) and (5.51),

$$\left.\begin{array}{l} N = 1 \cdot 276(600 + 563) + 563 = 2047 \text{ kN} \\ M_x = 1 \cdot 276(13 \cdot 4 + 14 \cdot 6) + 14 \cdot 6 = 50 \cdot 4 \text{ kNm} \end{array}\right\} \tag{5.52}$$

Since M_2 is entirely due to dead load, the ratio β due to the factored loads differs from that given by (5.36), but the effect of this can be neglected.

Failure about the x-axis. Using values from (5.46), (5.48), (5.50), and (5.52), $M_x/M_{px} = 0 \cdot 227$; and from (5.29),

$$N_x/N_u = 0 \cdot 84 - (0 \cdot 84 - 0 \cdot 36)(0 \cdot 227) = 0 \cdot 73 \tag{5.53}$$

From (5.39) and (5.52),

$$N/N_u = 0 \cdot 544 \tag{5.54}$$

Thus N_x exceeds the applied load N, and the column has a substantial margin of safety (about 34%) against failure about the x-axis only, as shown by points D and F in Fig. 5.8.

Biaxial bending.   In the present problem, $M_y = 0$, so from (5.29) and (5.48),

$$N_y/N_u = K_{1y} = 0.75 \tag{5.55}$$

From (5.11) and (5.48),

$$N_{ax}/N_u = K_{1x} = 0.84 \tag{5.56}$$

so from (5.30) and (5.53)

$$N_u/N_{xy} = 1/0.73 + 1/0.75 - 1/0.84 \tag{5.57}$$

whence

$$N_{xy}/N_u = 0.66 \tag{5.58}$$

This is about 9% below N_x/N_u, and shows that when this design method is used the effect of the initial curvature about the y-axis on the calculated failure load is found to be significant even when the imposed y-axis moments are zero. From (5.54) and (5.58), $N = 0.82\, N_{xy}$, so the column is also strong enough in biaxial bending.

Simplifications of the method.   The results of the preceding calculations are modified as follows if the proposals made in Section 5.3.4 are adopted. The former values are given in parentheses.

When f_u is reduced from $0.55 f_{cu}$ to $0.45 f_{cu}$, N_u becomes 3467 kN (3762 kN), and α is 0.37 (0.42). The change in N_u increases r_x from 74.6 to 78 mm, so L/r_x becomes 51.5 (53.6). If the creep factor m is taken as 1.0, the loads are $N = 1726$ kN (2047 kN) and $M_x = 42.7$ kNm (50.4 kNm), giving $N_x = 2600$ kN (2745 kN). So the load N is 66% (74%) of the strength N_x. The effect of omitting the creep factor outweighs the effect of the lower value of f_u, due mainly to the high proportion of load (67%) assumed in this example to be long-term.

5.3.7   Comparisons between column design methods

Account has to be taken of so many independent variables when a column length is designed that it is impossible to check fully any proposed design method by testing columns. Instead, the design loads given by simplified methods are compared with values computed by the most accurate method available.

For this purpose, 576 design loads were computed,[66] by using eight different methods for nine different combinations of axial load and bending about one or both axes, for composite columns of four cross-sections and two slenderness ratios. The methods included

(A) the Basu and Sommerville method (Section 5.3), which was assumed to be the most accurate available,
(B) the same method, but simplified as described in Section 5.3.4, and
(C) the cased-strut method of BS 449 (Section 5.2).

In the following account of some of the conclusions from this work, the symbols A, B, and C are used to represent the design loads at the ultimate limit state given by the methods (A), (B), and (C) for a given column length and given eccentricities of loading at the two ends and about the two axes of bending.

The simplified Basu method gave results that agreed with those from Method A to within $\pm 7\%$ except for slender columns with high concrete contribution factors (α). In these, the omission of creep coefficients and of the biaxial-bending calculation for columns loaded about the major axis only outweighed the effect of using reduced stresses in the concrete. For one column with $\alpha = 0.67$, $L/r_y = 150$, the ratio A/B ranged from 0.74 to 0.87 for the nine load cases. These differences are on the unsafe side. It is therefore recommended that method (B) is not used for column lengths for which both $\alpha > 0.4$ and $L/r_y > 100$, where r_y is calculated for the bare steel section.

The comparison between method (C) and method (A) gave interesting results. For short columns ($L/r_y = 40$) and for slender columns with concrete contribution factors exceeding 0.4, the cased-strut method was shown to be conservative, as expected, with many ratios A/C in the range 1.5 to 1.9. But for slender columns ($L/r_y = 150$) with $\alpha < 0.25$, several ratios A/C lay between 0.65 and 0.8. This is due to the difference between the Perry–Robertson strut curve (used in BS 449) and the curve K_1 shown in Fig. 5.3, which was discussed on p. 154. In view of the extensive experience of its use in practice, it is unlikely that the cased-strut method is as unsafe as these results imply. This suggests that at high slendernesses, the curve for K_1 is too low.

Due to the rule in the cased-strut method that the load on the column may not exceed twice that which would be permitted for the uncased section, some loads given by method (C) were less than half those from method (A). The purpose of the rule is to provide protection from the risk that due to lack of supervision or poor compaction, the concrete encasement may be weak or porous, particularly underneath the flanges of steel beams. A more rational way of allowing for weakness at beam–column joints would be to limit the axial load carried by the column to a 'reduced squash load', calculated on the assumption that only a certain proportion (e.g., one half) of the concrete encasement is present. It was found in the computations that this rule would govern design very rarely; it operated only for axially loaded short columns with concrete contribution factors exceeding about 0·4. If the real problem is thought to be weakness of the concrete all along the column length, rather than at the joints, the correct remedy is to increase γ_m for concrete encasement above the present value of 1·5, rather than to retain the present restrictive rule in the cased-strut method.

In conclusion, it can be said that the ultimate-strength method (A) usually shows that the steel section can be about one serial size smaller than that required by the cased-strut method (C). This saving is not sufficient to ensure that the method will be widely adopted unless the amount of calculation can be substantially reduced by the use of computer programs or simplifications of the kind suggested as method (B). The worst anomalies in the cased-strut method are likely to be removed in the version of it now being prepared for the revised BS 449, and this will reduce the savings of material that can be achieved by using methods (A) or (B).

5.4   Concrete-filled steel tubes

Where the fire risk to a structure is small, as in a bridge, columns made from concrete-filled tubes may be cheaper than concrete-encased I-sections or non-composite steelwork. A hollow steel section is more expensive than an open section of equal weight, but its bending strength and resistance to buckling are greater, due to the more efficient shape of the cross-section. No formwork or reinforcement are required for the concrete, and its surface is protected from impact and abrasion. For a given loading, a filled tube can usually be made more slender than an encased section, as is shown by the example in Section 5.4.2.

Information is now available on many of the problems that can arise in the design of filled tubes. In short 'pin-ended' columns, such as those used between levels of the four-level motorway interchange at Almondsbury,[6] the whole of the load is applied to the column at one point, unlike the situation in a building. It is then difficult to ensure that the load is so distributed that the compressive strains in the steel and the concrete are equal. In an early design for the Almondsbury columns, the load was applied only to the concrete filling, and tests found[67] that breakdown of bond occurred near the ends of the tube. In the final design, load was applied through thick steel endplates which bore both on the concrete and on the steel, and stud shear-connectors were provided on the inner surface of the tube near its ends. The problem was unusually severe in these large tubes, which were up to 42 in (1·07 m) in diameter. When standard tubes are used (circular sections up to 0·46 m in diameter, and square and rectangular sections of similar size), simple end-plates can be used, and shear connectors are not usually required. A useful set of typical joint designs and fabrication details is available in CIDECT Monograph No. 1.[65]

In the construction of filled tubes, it is essential to ensure that the concrete is properly compacted, for no subsequent inspection is possible, and to provide holes for drainage and for the relief of excessive vapour pressure that may build up during a fire. Information on these points and on the protection of filled tubes from fire is also given in Ref. 65.

5.4.1   Strength of filled tubes

A method of computing the load–deflection curve and the maximum load for an eccentrically-loaded concrete-filled tube has been developed at Imperial College[68] and compared with the results of tests on columns of circular cross-section. It was assumed in the theory that plane sections remain plane and that the uniaxial stress–strain curves for steel and concrete are applicable. Good agreement between theory and tests was found for columns with length–diameter ratios exceeding about 15, but stocky cylindrical columns were found to be stronger than the theory predicted.

At low stresses, concrete has a lower Poisson's ratio than steel, so there is a tendency for separation to occur between the two materials in a filled tube. At high compressive stresses, internal microcracking in concrete causes it to swell. Its outwards movement is restrained by the steel, and the increase in the strength of the concrete due to this lateral restraint more than outweighs the loss of compressive strength in the steel due to the circumferential tension. The increase in the strength of the column due to this effect has been studied by Furlong[69] and by Sen.[70] It decreases with increasing slenderness and increasing eccentricity of loading, and is small in square and rectangular cross-sections, so it cannot often be exploited in practice. No account is taken of it in the design methods discussed below.

The computer program developed by Neogi, Sen, and Chapman[68] has been used to calculate ultimate loads for pin-ended filled tubes made from the current British range of circular, square, and rectangular hollow sections, and the results have been published as design tables.[65] There are three main reasons why this method may give different results from those obtained by the method of Basu and Sommerville (Section 5.3):

(1) For bending about each axis, the greater end moment is assumed to be applied at both ends of the column length, so as to cause single-curvature bending (i.e., β is taken as $+1$). The method therefore underestimates the strength of columns subjected to double-curvature bending, particularly at high L/r ratios.

(2) The modification factor for creep of concrete (p. 161) is, in effect, applied to the whole load on the column, rather than to the long-term load only.

(3) In developing their method, Basu and Sommerville used formulae which gave lower bounds to their computed results, and so may be well on the safe side for particular columns; but the tabulated figures are as computed for particular columns, and for this reason sometimes give higher design loads.

The first two of these three differences between the methods operate in the opposite direction to the third, and their relative weights vary from column to column; so all that can be said is that the difference between results given by the two methods should be small, as it is in the example given below.

When the design tables were prepared, it was not known what partial safety factors for materials (γ_m) would be adopted for limit-state design, so the tables are based on design strengths for both steel and concrete, not characteristic strengths, with the result that data are given for values different from those likely to be used in practice. If a revised edition of the book is issued in metric units and in terms of characteristic strengths, ultimate-strength design of pin-ended filled tubes will become a rapid and straightforward process.

It is assumed in the following example that the column length designed in Section 5.3.6 is to be replaced by a concrete-filled tube of square cross-section. The CIDECT tables[65] are used to find an appropriate section, and its strength is then calculated by the method of Basu and Sommerville.

An expression for the moment of resistance of a rectangular filled tube is now derived. The method is similar to that used in Section 5.3.1 for an encased section.

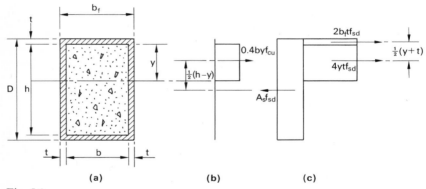

Fig. 5.9

The cross-section and dimensions are shown in Fig. 5.9(a), and the longitudinal forces in Figs. 5.9(b) and 5.9(c). Resolving longitudinally,

$$0\cdot4byf_{cu} + (2b_f t + 4yt)f_{sd} = A_s f_{sd} \tag{5.59}$$

whence

$$y = (A_s - 2b_f t)/(4t + 0\cdot4bf_{cu}/f_{sd}) \tag{5.60}$$

Taking moments,

$$M_p = [\tfrac{1}{2}A_s(h - y) + b_f t(y + t)]\, f_{sd} \tag{5.61}$$

5.4.2 Example (continued). Use of filled tube for external column

To enable the reader to follow the way in which Ref. 65 is being used, part of the following calculation is given in Imperial units. In the previous calculation (Section 5.3.6) a column 32 cm square was found to be suitable. Knowing that filled tubes are more compact than encased sections, we assume that a square hollow section of side 25·4 cm (10 in) will be used. The reduction in self-weight is negligible, but account will be taken of the reduction in the eccentricity of the load from the beam, from 26 cm to $\frac{1}{2} \times 25·4 + 10 = 22·7$ cm. From (5.34), the new design moments are

$$M_g = 11·7 \text{ kNm} \qquad M_q = 25·6 \text{ kNm} \qquad (5.62)$$

The axial load is 1726 kN as before (173 tons), so the eccentricity of loading (e/D in Ref. 42) when $D = 0·254$ m is

$$(M_d + M_i)/ND = 37·3/(1726 \times 0·254) = 0·085 \qquad (5.63)$$

The effective length is 4 m as before (13 ft), and the design stresses f_{cu}/γ_m and f_{sd} are 20 N/mm^2 (2900 lb/in^2) and 304 N/mm^2 (19·7 ton/in^2). Table T 64 of Ref. 65 indicates that a $10 \times 10 \times 0·250$ inch section may be strong enough. To obtain its failure load, interpolation is necessary between tabulated values for f_{cu}, f_{sd}, e/D, and effective length, and the result is

$$N_x = 186 \text{ tons} \qquad (5.64)$$

We assume equal lateral restraint about both axes. The member is equally stiff about both axes, so y-axis buckling cannot occur, and (5.64) gives the required failure load. The section is therefore suitable, and is loaded to 93% ($100 \times 173/186$) of its calculated strength.

The strength of this column is now analysed by the method of Basu and Sommerville. Its dimensions and properties are shown in Fig. 5.10. From (5.9),

$$N_u = 6·27 \times 304 + 58·1 \times 16·6 = 1906 + 964 = 2870 \text{ kN} \qquad (5.65)$$

From (5.13),

$$\alpha = 96/2870 = 0·336 \qquad (5.66)$$

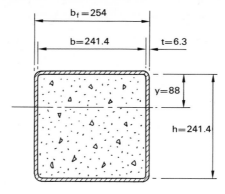

Fig. 5.10

From (5.12),

$$r^2 = (250 \times 6410 + 0\cdot24 \times 30 \times 28\ 100)/287 = 6290\ \text{mm}^2$$

so that

$$r_x = r_y = 79\ \text{mm} \qquad L/r = 51 \qquad (5.67)$$

From (5.60),

$$y = \frac{10(62\cdot7 - 50\cdot8 \times 0\cdot63)}{2\cdot52 + 12 \times 24\cdot14/304} = 88\ \text{mm}$$

From (5.61),

$$M_p = (31\cdot4 \times 153 + 25\cdot4 \times 0\cdot63 \times 94)304 \times 10^{-4} = 192\ \text{kNm} \tag{5.68}$$

From (5.31),

$$m = 1 + 3\alpha/8 = 1\cdot126$$

Assuming, as on p. 168, that half of the imposed load is long-term, the design loads are

$$\left.\begin{aligned} N &= 1\cdot13(600 + 563) + 563 = 1873\ \text{kN} \\ M_x &= 1\cdot13(11\cdot7 + 12\cdot8) + 12\cdot8 = 40\cdot3\ \text{kNm} \end{aligned}\right\} \qquad (5.69)$$

From (5.24),

$$e_x = 40\cdot3/1873 = 0\cdot0215\ \text{m}$$

From (5.25),

$$\eta = 2870/192 = 15{\cdot}0\ \mathrm{m}^{-1}$$

so that

$$(\eta e)_x = 0{\cdot}322 \qquad (5.70)$$

From (5.19),

$$K_{20} = 0{\cdot}9 \times 0{\cdot}336^2 + 0{\cdot}2 = 0{\cdot}302 \qquad (5.71)$$

The ratio of end moments is as before, so from (5.36), $\beta = -0{\cdot}314$. Substitution in (5.18) gives $(K_2/K_{20}) > 1$, so the ratio is taken as $1{\cdot}0$, and from (5.71),

$$K_2 = 0{\cdot}302$$

From Fig. 5.3 with $L/r = 51$,

$$K_1 = 0{\cdot}85$$

From (5.27),

$$\frac{N_x}{N_u} = \frac{0{\cdot}85}{1 + 0{\cdot}548 \times 0{\cdot}322} = 0{\cdot}723 \qquad (5.72)$$

Biaxial bending is now considered. For a square filled tube with $M_y = 0$, $N_y = N_{ay} = N_{ax}$, so (5.30) gives $N_{xy} = N_x$, and (5.72) therefore gives the required ultimate load:

$$N_x = 0{\cdot}723 \times 2870 = 2074\ \mathrm{kN} \qquad (5.73)$$

From (5.69), the design is satisfactory, and the column is loaded to 90% of its calculated strength. The result is in close agreement with that given by the CIDECT Tables (93%).

5.5 Beam–column joints and the design of composite frames

The behaviour of a composite frame at the ultimate limit state is strongly influenced by the design of the joints between the beams and the columns. 'Simple' design as used in Section 5.2 is based on the assumption that the column is continuous past the joint, and that the beams are pin-jointed to its faces. The design of the columns may be unsafe unless it is possible for the end of each beam to rotate relative to the column with negligible transfer of bending moment between the two.

There has been no systematic research on the moment–rotation characteristics of 'simple' joints between composite beams and columns. The largest transfers of bending moment are likely to occur at external columns and at internal columns supporting beams of

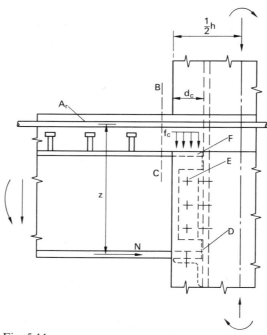

Fig. 5.11

unequal depth. Study of a typical joint (Fig. 5.11) subjected to negative moment suggests that the necessary conditions for the simple design method to be valid are:

(1) The top longitudinal reinforcement in the slab (A_r) consists only of small-diameter bars, as required to control cracking of the slab on plane BC and to hold the transverse reinforcement in place.
(2) The steel beam shall have no end-plate or top cleat, and the bolts to the web cleat shall not be close-tolerance or friction-grip bolts.

This is necessary because there may be no clearance at point D, so that the beam must rotate about this point, with slip of the bolts at E.

(3) Increased eccentricities of load should be specified for 'simple' joints in which d_c exceeds, say, 70 mm, as it would in joints between beams and the webs of encased I-sections. The reason is that outwards movement of the flange at F will be resisted by friction due to the compressive stress f_c in the encasement to the column. It is shown in Section 5.5.1 that this may cause moments greater than those given by the use of the nominal eccentricities specified in BS 449.

The encasement of the composite beam in concrete should make little difference to moment transfer at the joint, provided that no reinforcement crosses plane BC (Fig. 5.11), because the crack in the slab on this plane is likely to extend into the encasement, and the horizontal compressive force at D could not increase unless there was an equal increase in tensile force higher up.

If these conditions are not satisfied, the obvious alternative is to analyse the frame as if rigid jointed, or to use one of the simplified procedures based on elastic theory that have been developed for reinforced concrete frames.[9, 60] It is then necessary to ensure that the beam–column joints are capable of resisting the large bending moments given by the analysis. This is likely to involve providing the steel beams with end-plates and attaching them to the steel columns by welding or friction-grip bolting. Such joints are expensive, partly because it is difficult to make provision in multi-bay frames for the inevitable variations in the lengths of steel beams and the depths of column sections.

There is another reason why rigid-jointed composite frames de-signed on an elastic basis are likely to be uneconomic in buildings. It was shown in Chapter 4 that bending-moment distributions in com-posite beams given by elastic theory differ substantially from the distribution of available strength. The same is true of rigid-jointed frames, and it follows that in buildings, plastic design of beams should be used wherever possible. If the beams are made fully continuous, the limiting web and compression-flange slendernesses for plastic design (Section 4.4.2) are more onerous than for steel frames, and rule out the use of the more slender sections unless local stiffening is provided.

A possible alternative to rigid joints is to use an inexpensive beam–column joint similar to that shown in Fig. 5.11, but to provide quite heavy longitudinal reinforcement in the slab. If d_c is small enough for the horizontal force due to the stress f_c to be negligible, the longitudinal compressive force in the bottom flange of the steel beam, N, can be assumed to be $A_r f_{rd}$ at flexural failure of the beam, and A_r can be so chosen that this force is not enough to cause the flange or web of the steel section to buckle. The negative moment of resistance of the beam can be assumed to be $A_r f_{rd} z$. The design is efficient partly because z, the lever arm, is a high proportion of the total depth available.

In research on joints of this type,[71] their moment–rotation curves have been compared with those of rigid joints between beams of similar slenderness and a column. Typical results are shown in Fig. 5.12, in which θ is the rotation of the point of contraflexure in the beam relative to the column. HB 41 is a rigid-jointed composite beam with a

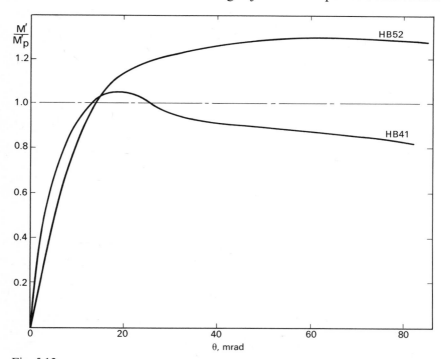

Fig. 5.12

web slenderness (d_w/t_w) of 40·5, yield stress 330 N/mm^2, and a force ratio of 0·27. According to inequality (4.18), it is just too slender for use in plastic design, and its M–θ curve shows clearly the loss of strength after local buckling. HB 52 is a composite beam with a semi-rigid joint and the properties $d_w/t_w = 46$, $f_y = 277$ N/mm^2, and $\phi = 0·35$. It does not satisfy (4.18), but its M–θ curve shows that the joint has excellent rotation capacity without loss of strength, so that its use would enable the beam to be designed by plastic theory.

In a beam with semi-rigid joints, the cracks in the top surface of the slab near the columns are likely to be wider at working load than when rigid joints are used. In many buildings, this will not matter, for the widely used design method in which steel beams are assumed to be simply supported also implies that large tensile strains will occur at floor level near internal columns, and in practice this is not found to cause unserviceability.

The case for using semi-rigid joints is not primarily that they increase the strength of the beam, because additional strength can usually be provided more cheaply at midspan. Their advantages are that the addition of top longitudinal reinforcement at the supports is often the cheapest way of reducing an excessive midspan deflection at working load, and that when 'simple' design is inapplicable, they can be used to provide continuity at a cost lower than that of rigid joints.

5.5.1 Example (continued). Fixing moments in 'simple' joints

It is assumed that the beam section shown in Fig. 3.7 is attached to an external column of the section shown in Fig. 5.1 by a 'simple' joint as sketched in Fig. 5.11. The beam is designed as simply supported. When loaded, it rotates relative to the column about point D in Fig. 5.11, causing slip at point F and yielding of the anti-crack reinforcement A_r. The moment imposed on the column will now be estimated.

For the slab, A_r is secondary reinforcement, and CP 110 states that it should not be less than 0·15% of the area of the slab. This rule gives 8-mm mild-steel bars at about 0·24 m pitch. It is likely that three of these would be anchored in the column encasement, so A_r is taken as 1·51 cm^2, and $A_r f_{rd} z$ (Fig. 5.11) is about 20 kNm if z is taken as 52 cm.

At the ultimate limit state, the mean compressive stress in the concrete encasement is likely to be at least 10 N/mm². The area of beam flange embedded in this concrete is $5.7 \times 15.3 = 87$ cm². Assuming a friction coefficient of 0·4 and allowing for two surfaces of sliding at F in Fig. 5.11, the horizontal force is $8.7 \times 10 \times 0.4 \times 2 = 70$ kN. The lever arm to the bottom flange is 0·4 m, giving a moment of 28 kNm.

If the end reaction for this beam, 330 kN (Section 5.3.6) is assumed to act through the bolts to the web cleat, the eccentricity is about 0·14 m, giving a further moment of 46 kNm, and a total for the three effects of 94 kNm. With a connection to the web of the stanchion, similar calculations give the imposed moment as 143 kNm. The proposed design eccentricity (Section 5.3.6) is 0·26 m for both axes, giving the moment applied to the column as 86 kNm. This is reasonably close to the estimated major-axis moment, but well below the minor-axis value of 143 kNm. The true value would be higher still due to the fixity provided by the bolted joints. These figures give some idea of the approximate nature of the 'simple' design method.

5.6 Design of rigid-jointed composite frames

Columns have little resistance to lateral forces, so in framed structures it is usual to provide walls or diagonal bracing to resist loading due to wind and other horizontal forces. The sidesway of the structure is then so small that its effect on the behaviour of the columns is negligible. Design methods for such 'braced' columns are applicable in structures of any height.

When the columns are assumed to provide resistance to lateral forces, they are known as 'unbraced'. In frames more than about three storeys high, account must be taken of the eccentricity of loads on the columns due to sidesway of the frame. No simple method of analysis that takes account of the additional sidesway due to inelastic behaviour of the columns is yet available, so the frame should be analysed and the columns designed according to elastic theory. The method given in CP 110 for concrete frames could be adapted for use with a composite rigid-jointed frame. It is not described here, because tall unbraced frames are rarely used in practice.

It is possible for a column to be 'braced' in one plane and 'unbraced' in the other, but in most frames the columns are braced about both axes, and only these are now considered.

5.6.1   Braced frames

When the beams are designed as simply supported and the beam–column joints are 'simple' as discussed in Section 5.5, the columns can be designed by the methods of Section 5.2 or 5.3, assuming in both methods that the effective length of the column is not less than the actual length between points at which lateral restraint is provided.

In rigid-jointed braced frames, these methods are conservative for beams and perhaps unsafe for columns, as explained above. Research on a better method is in progress. A summary is now given of the present state of this work, and an interim design method is proposed.

It is assumed first that individual spans of all beams are designed by simple plastic theory, assuming hinges to occur at each end and in the midspan region. Where it is expected that the available restraint from the column may be less than the resistance moment of the beam (for example, where an external column supports a long-span roof beam), the end fixity should be based on a preliminary design for the column.

The most critical arrangement of imposed loading for a column length can be assumed to be when the axial load in the length above it is the maximum possible, and the beams framing into the ends of the length considered are so loaded as to cause the maximum degree of single-curvature bending about the minor axis, or, if minor-axis beam loads are low, about the major axis.

For example, if it is now assumed that the internal support for the two-span beams of the worked example of Chapter 4 is a column and not a wall, the loading for the design of column length *AB* in Fig.

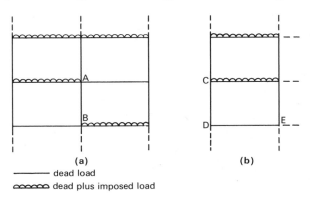

(a)   (b)

————— dead load

ᴖᴖᴖᴖᴖ dead plus imposed load

Fig. 5.13

5.13(a) is as shown, with all beams above level *A* carrying full imposed load. Similarly, the loading for the external column *CD* is shown in Fig. 5.13(b). This loading causes double-curvature bending, but to a smaller extent than if beam *DE* also carried imposed load. The diagrams show the major-axis beams because in this frame they are sufficiently closely spaced for loads on the minor-axis beams, if any, to be small.

5.6.2   Beam–column interaction

The nature of the interaction between beams and columns is the heart of the problem, for, as has been shown, failure loads of individual beams and columns can be found with acceptable accuracy. At present, the major-axis and minor-axis planes are considered separately, and the braced frame in each plane is represented by a limited frame,[48] consisting of the member being studied and those attached to its ends. The other ends of the attached members are assumed to be fixed. Fully loaded beams are assumed to have plastic hinges at their ends, and so to impose a constant moment M_p on the joint, irrespective of the joint rotation, θ.

In the design of rigid-jointed steel frames,[48] beams carrying dead load only are assumed to be elastic. The use of this assumption in

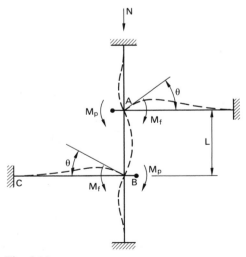

Fig. 5.14

composite frames is now considered. The limited frame corresponding to Fig. 5.13(a) is then as shown in Fig. 5.14. If M_f is the fixed-end moment of the unloaded beams, the unbalanced moment at each end of the column when $\theta = 0$ is $M_p - M_f$. The effect of these moments, of initial curvature of the column length, and of increasing axial load N is that the joints A and B rotate as shown in Fig. 5.14. If the moment–rotation curves for the beams were bi-linear elastic–plastic, and if the beams either side of joint B were of equal strength, then the relation between the clockwise rotation of joint B and the net clockwise moment on it from the beams would be as in Fig. 5.15. In the Figure,

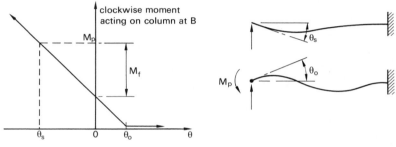

Fig. 5.15

θ_s is the rotation at which the end moment in the elastic beam BC is zero, and θ_0 is that at which it reaches M_p. For $\theta > \theta_0$, the net moment on the column is zero. This shows that the column length AB could be designed for end moments less than $M_p - M_f$, and that if its end rotation exceeded θ_0 before its maximum axial load were reached, then the end moments could be taken as zero, and the ultimate load would be that for a pin-ended column of length L.

Thus, beam–column interaction could be allowed for by reducing the out-of-balance end moment but not the effective length. To work out this idea in detail requires knowledge of the true M–θ curves for the ends of beams and the relation between axial load and end rotation for columns. Research on these subjects is in progress.

5.6.3   Critical buckling and effective length

Another approach to the problem of beam–column interaction is via the elastic critical buckling load, N_{cr}, for the column length as restrained by the beams in the limited frame. This can easily be found from tabulated stability functions if the unloaded beams and the column are assumed to remain elastic. When appropriate account is taken of the deterioration of stiffness of the column due to inelastic behaviour, a deteriorated critical load can be found, which is the ultimate load of the column if the restraining beams are still elastic. This forms the basis of Wood's simplified method for the design of steel columns.[72] There are two obstacles to its use for composite columns. The first is that simplified expressions for the deteriorated stiffness of composite columns have yet to be developed. The second is a consequence of the extensive redistribution of moment required in continuous composite beams, discussed in Chapter 4. When $\theta = 0$, the negative-moment region of a span carrying factored dead load only may already be at or near the limit of elastic behaviour. It may be unable to accept its share of the out-of-balance moment at the joint, assumed to be transferred to it by rotation of the joint, without deterioration of stiffness. A method of assessing this deterioration and of allowing for it is not yet available.

In a braced frame with elastic beams, the elastic critical buckling load N_{cr} for a column of length l exceeds its Euler buckling load N_e (which is $\pi^2 EI/l^2$), so that if the effective column length L is defined by

$$N_{cr} = \pi^2 EI/L^2 \qquad (5.74)$$

then $L < l$. The use of effective-length ratios L/l less than one is the way in which restraint from beams is allowed for in the 'simple' column design method of BS 449.

In the draft Bridge Code[15] it has been proposed that effective-length ratios calculated from undeteriorated stiffnesses can be used in the design of braced frames. The use of the full stiffness of beams is reasonable in bridges, since these beams are unlikely to be designed on a plastic basis. The method is appropriate when columns are designed elastically, for it then corresponds with that for multi-storey welded steel frames;[48] but no detailed justification is yet available for its use in conjunction with the Basu and Sommerville ultimate-strength method for columns, as is proposed.

Thus some existing methods allow for beam–column interaction by reducing the effective length of the column and sharing the out-of-balance end moment between the column and the beams that carry dead load only. These methods may not be applicable to composite frames with beams designed plastically, for the reasons given above.

5.6.4   Interim design method

This method is based on the conservative assumption that the beams carrying factored dead load only have hinges at their ends but not at midspan. They can accept additional moment due to transverse loading, but not due to rotation of the beam–column joint. It follows that the effective-length ratio L/l may not be reduced below one due to restraint from composite beams, and that the whole out-of-balance end moment must be shared between the upper and lower column lengths. The shares should strictly be in proportion to the deteriorated stiffnesses. Equal shares are accurate enough unless there is a sudden change of cross-section, when shares in proportion to the elastic stiffnesses can be used.

These assumptions are not as conservative as they would be for steel beams. For internal joints the out-of-balance moment is the difference between M'_p and the dead-load fixed-end moment given by elastic theory, and this difference becomes smaller as the ratio of midspan hinge moment to support hinge moment increases.

Each column length may be designed by the ultimate-strength method to resist the axial load and the end moments about each axis found in this way. The method is likely to be modified by the results of research now in progress, so it has been presented here in outline only. An example of its use is now given.

5.6.5   Example (continued). Design of rigid-jointed braced frame

The problem and the loads are assumed to be as before. An external column length supporting five storeys will be designed for the ultimate limit state. Rigid joints lead to economy in beams, not in columns, and might be used if it was important to have beams of minimum depth, as was the case in a rigid-jointed composite frame built in 1963–4.[7] A typical 10·5-m span beam must therefore be redesigned.

From Table 4.1, the dead load is 22 kN/m, and the total load is 70 kN/m, giving $wL^2/8 = 965$ kNm. The relevant beam loading and the limited frame for the design of column length AC are shown in Fig. 5.16. The design condition for a typical beam is

$$M_p + M'_p \not< 965 \text{ kNm} \qquad (5.75)$$

The cross-section will be as in Fig. 3.7, but with a smaller steel section. A Universal Beam 359 × 172 mm (57 kg/m) would probably be strong enough, but its flange slenderness ratio (13·2) is outside the limit given in (4.17) for Grade 50 steel. So a section 385 × 153 mm (60 kg/m) is chosen. This has $b/t_f = 10·6$, which is acceptable. For the midspan cross-section, $M_p = 667$ kNm, and the plastic moment of the steel section alone is 336 kNm, so (5.75) is satisfied even if the slab reinforcement in the negative moment region at B is neglected.

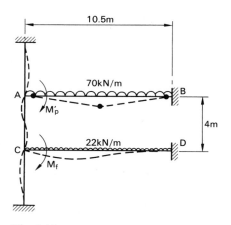

Fig. 5.16

The vertical shear at A is $5·25 \times 70 = 368$ kN, and if the column self-weight is 84 kN as before (p. 165), the design axial load is

$$N = 5 \times 368 + 84 = 1922 \text{ kN} \qquad (5.76)$$

Assuming the beam moments to be shared equally between upper and lower column lengths, the design moments for length AC are

$$M_1 = \tfrac{1}{2} \times 336 = 168 \text{ kNm} \qquad (5.77)$$
$$M_2 = \tfrac{1}{2}(gL^2/12) = 101 \text{ kNm} \qquad (5.78)$$

The simplifications proposed in Section 5.3.4 are adopted, so that magnification factors are taken as 1·0, and from (5.77) and (5.78),

$$\beta_x = -101/168 = -0\cdot6 \tag{5.79}$$

This loading is significantly higher than that for which the encased section of Fig. 5.1 was designed in Section 5.3.6, but it was then found to have ample strength, and so will be checked for the new loading.

K_{1x}, K_{20}, and K_{2x} are as before, and $e_x = 168/1922 = 0\cdot0874$ m. As before, $\eta_x = 17\cdot0$ m^{-1}, so $(\eta e)_x = 1\cdot485$. From (5.27),

$$N/N_u = 0\cdot85/(1 + 0\cdot526 \times 1\cdot485) = 0\cdot477 \tag{5.80}$$

When f_u is taken as $0\cdot45f_{cu}$, then $N_u = 3467$ kN (p. 169), so from (5.80),

$$N \not> 0\cdot477 \times 3467 = 1654 \text{ kN}$$

From (5.76), this section is not strong enough, and a heavier 8 × 8 in steel section and/or more longitudinal reinforcement is required. The axial load on this external column is increased by 11% due to the change in the shape of the bending-moment distribution in the beams, so there would be a corresponding reduction in the design load for the internal column or wall.

Thus the use of rigid beam–column joints in this example gives lighter and shallower beams and heavier external columns; it may enable weaker internal columns to be used, due to the reduced axial load.

Appendix A. Partial-interaction Theory

A.1 Theory for simply supported beam

This subject is introduced in Section 2.6 (p. 33), which gives the assumptions and notation used in the theory that follows. On first reading, it may be found helpful to re-write the algebraic work in a form applicable to a beam with the very simple cross-section shown in Fig. 2.2. This can be done by making these substitutions:

Replace A_c and A_s by bh, and d_c by h.
Replace I_c and I_s by $bh^3/12$.
Put $k_c = m = 1$, so that E_c', E_c, and E_s are replaced by E.

The beam to be analysed is shown in Fig. 2.14, and Fig. A.1 shows in elevation a short element of the beam, of length dx, distant x from the midspan cross-section. For clarity, the two components are shown separated, and displacements are much exaggerated. The slip is s at cross-section x, and increases over the length of the element to $s + (ds/dx) \, dx$, which is written as s^+. This notation is used in Fig. A.1 for increments in the other variables, M_c, M_s, F, V_c, and V_s, which are respectively the bending moments, axial force, and vertical shears acting on the two components of the beam, the suffixes c and s indicating concrete and steel. It follows from longitudinal equilibrium that the forces F in steel and concrete are equal. The interface vertical force r per unit length is unknown, so it cannot be assumed that V_c equals V_s.

If the interface longitudinal shear is q per unit length, the force on each component is $q\,dx$. It must be in the direction shown, to be consistent with the sign of the slip, s. The load–slip relationship is

$$pq = ks \qquad (A.1)$$

since the load per connector is pq.

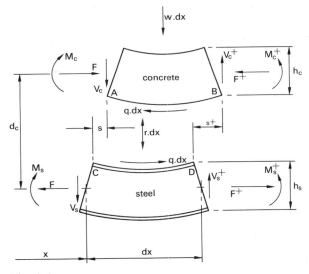

Fig. A.1

We first obtain equations deduced from equilibrium, elasticity, and compatibility, then eliminate M, F, V, and q from them to obtain a differential equation relating s to x, and finally solve this equation and insert the boundary conditions. These are

(1) Zero slip at midspan, from symmetry, so

$$s = 0 \qquad \text{when} \qquad x = 0 \qquad (A.2)$$

(2) At the supports, M and F are zero, so the difference between the longitudinal strains at the interface is the differential strain, ε_c, and therefore

$$ds/dx = -\varepsilon_c \qquad \text{when} \qquad x = \pm L/2 \qquad (A.3)$$

Equilibrium

Resolve longitudinally for one component:

$$dF/dx = -q \tag{A.4}$$

Take moments:

$$\frac{dM_c}{dx} + V_c = \tfrac{1}{2}qh_c \qquad \frac{dM_s}{dx} + V_s = \tfrac{1}{2}qh_s \tag{A.5}$$

The vertical shear at section x is wx, so

$$V_c + V_s = wx \tag{A.6}$$

Now $\tfrac{1}{2}(h_c + h_s) = d_c$, so from (A.5) and (A.6),

$$\frac{dM_c}{dx} + \frac{dM_s}{dx} + wx = qd_c \tag{A.7}$$

Elasticity

In beams with adequate shear connection, the effects of uplift are negligible in the elastic range. If there is no gap between the two components, they must have the same curvature, ϕ, and simple beam theory gives the moment–curvature relations. Using (2.15) for E_c', then

$$\phi = \frac{M_s}{E_s I_s} = \frac{m M_c}{k_c E_s I_c} \tag{A.8}$$

The longitudinal strains in concrete along AB (Fig. A.1) and in steel along CD are:

$$\varepsilon_{AB} = \tfrac{1}{2}h_c\phi - \frac{mF}{k_c E_s A_c} - \varepsilon_c \tag{A.9}$$

$$\varepsilon_{CD} = -\tfrac{1}{2}h_s\phi + \frac{F}{E_s A_s} \tag{A.10}$$

Compatibility

The difference between ε_{AB} and ε_{CD} is the slip strain, so from (A.9) and (A.10), and putting $\tfrac{1}{2}(h_c + h_s) = d_c$,

$$\frac{ds}{dx} = \phi d_c - \frac{F}{E_s}\left(\frac{m}{k_c A_c} + \frac{1}{A_s}\right) - \varepsilon_c \tag{A.11}$$

It is now possible to derive the differential equation for s. Eliminating M_c and M_s from (A.7) and (A.8),

$$E_s\left(\frac{k_c I_c}{m} + I_s\right)\frac{d\phi}{dx} + wx = qd_c \tag{A.12}$$

From (A.1) and (2.18),

$$\frac{d\phi}{dx} = \frac{kd_c s/p - wx}{E_s I_0} \tag{A.13}$$

Differentiating (A.11) and eliminating ϕ from (A.13), F from (A.4), and q from (A.1):

$$\frac{d^2 s}{dx^2} = \frac{kd_c^2 s/p - wd_c x}{E_s I_0} + \frac{ks}{E_s A_0 p} = \frac{ks}{pE_s I_0}\left(d_c^2 + \frac{I_0}{A_0}\right) - \frac{wd_c x}{E_s I_0}$$

Introducing A' from (2.17), α^2 from (2.19) and β from (2.20) gives result (2.21), which is in a standard form:

$$\frac{d^2 s}{dx^2} - \alpha^2 s = -\alpha^2 \beta wx \tag{2.21}$$

Solving for s,

$$s = K_1 \sinh \alpha x + K_2 \cosh \alpha x + \beta wx \tag{A.14}$$

The boundary conditions (A.2) and (A.3) give

$$K_2 = 0 \qquad \varepsilon_c = -K_1 \alpha \cosh(\alpha L/2) - \beta w$$

and substitution in (A.14) gives s in terms of x:

$$s = \beta wx - \left(\frac{\beta w + \varepsilon_c}{\alpha}\right)\operatorname{sech}\left(\frac{\alpha L}{2}\right)\sinh \alpha x \tag{2.23}$$

Other results can now be found as required. For example, the slip strain at midspan is

$$\left(\frac{ds}{dx}\right)_{x=0} = \beta w - (\beta w + \varepsilon_c)\operatorname{sech}(\alpha L/2) \tag{A.15}$$

and the slip at $x = L/2$ due to ε_c alone (i.e., with $w = 0$), is

$$(s)_{x=L/2} = -(\varepsilon_c/\alpha)\tanh(\alpha L/2) \tag{A.16}$$

A.2   Example. Partial interaction

These calculations are introduced in Section 2.7. They relate to a beam shown in section in Fig. 2.15, which carries a distributed load w per unit length over a simply supported span L. The materials are assumed to be concrete with a characteristic cube strength of 30 N/mm² and mild steel, with a characteristic yield strength of 250 N/mm². Creep is neglected ($k_c = 1$) and we assume $m = 10$, so for the concrete $E_c = E'_c = 20$ N/mm², from (2.15).

The dimensions of the beam (Fig. 2.15) are so chosen that the transformed cross-section is square: $L = 10$ m, $b = 0.6$ m, $h_c = h_s = 0.3$ m. The steel member is thus a rectangle of breadth 0.06 m, so that $A_s = 0.018$ m², $I_s = 1.35 \times 10^{-4}$ m⁴.

The design of such a beam on an ultimate-strength basis is likely to lead to a working or 'service' load of about 35 kN/m. If stud connectors 19 mm in diameter and 100 mm long are used in a single row, an appropriate spacing would be 0.18 m. Push-out tests give the ultimate shear strength of such a connector as about 100 kN, and the slip at half this load is usually between 0.2 and 0.4 mm. Connectors are found to be stiffer in beams than in push-out tests, so a connector modulus $k = 150$ kN/mm will be assumed here, corresponding to a slip of 0.33 mm at a load of 50 kN per connector.

The distribution of slip along the beam and the stresses and curvature at midspan are now found by partial-interaction theory, using the results obtained in Section A.1, and also by full-interaction theory. The results are discussed in Section 2.7 (p. 36).

First α and β are calculated. From (2.18) with $I_c = mI_s$ (from the shape of the transformed section) and $k_c = 1$, $I_0 = 2.7 \times 10^{-4}$ m⁴. From (2.16) with $A_c = mA_s$ and $k_c = 1$, $A_0 = 0.009$ m². From (2.17), $1/A' = 0.3^2 + (2.7 \times 10^{-4})/0.009 = 0.12$ m². From (2.19), with $k = 150$ kN/mm and $p = 0.18$ m,

$$\alpha^2 = \frac{150 \times 0.12}{0.18 \times 200 \times 0.27} = 1.85 \text{ m}^{-2}$$

whence $\alpha = 1.36$ m⁻¹. Now $L = 10$ m, so $\alpha L/2 = 6.8$, and sech $(\alpha L/2) = 0.00223$. From (2.20),

$$\beta = \frac{0.18 \times 0.3}{0.12 \times 150 \times 1000} = 3.0 \times 10^{-6} \text{ m/kN}$$

We assumed $w = 35$ kN/m, so $\beta w = 1.05 \times 10^{-4}$ and $\beta w/\alpha = 0.772 \times 10^{-4}$ m. An expression for the slip in terms of x is now given by (2.23) with $\varepsilon_c = 0$:

$$10^4 s = 1.05x - 0.0017 \sinh (1.36x) \qquad (2.24)$$

This gives the maximum slip (when $x = \pm 5$ m) as ± 0.45 mm.

This may be compared with the maximum slip if there were no shear connection, which is given by (2.6) as

$$\frac{wL^3}{4Ebh^2} = \frac{35 \times 10^3}{4 \times 20 \times 0.6 \times 0.3^2 \times 1000} = 8.1 \text{ mm}$$

The stresses at midspan can be deduced from the slip strain and the curvature. Differentiating (2.24) and putting $x = 0$,

$$10^4 (ds/dx)_{x=0} = 1.05 - 0.0017 \times 1.36 = 1.05$$

so the slip strain at midspan is 105×10^{-6}. From (A.13),

$$(d\phi/dx) = 4.64s - 6.5 \times 10^{-4}x$$

Using (2.24) for s and integrating,

$$10^6 \phi = -81.5x^2 - 0.585 \cosh (1.36x) + K$$

The constant K is found by putting $\phi = 0$ when $x = L/2$, whence at $x = 0$,

$$\phi = 0.0023 \text{ m}^{-1}$$

The corresponding change of strain between the top and bottom faces of a member 0.3 m deep is 0.3×0.0023, or 690×10^{-6}. The transformed cross-section is symmetrical about the interface, so the strain in each material at this level is half the slip strain, say 52×10^{-6}, and the strain distribution is as shown in Fig. 2.16. The stresses in the concrete, found by multiplying the strains by E_c (20 kN/mm^2), are 1.04 N/mm^2 tension and 12.8 N/mm^2 compression. The tensile stress is below the cracking stress, as assumed in the analysis.

The maximum compressive stress in the concrete is given by full-interaction theory (Eq. 2.7) as

$$f_{cf} = \frac{3wL^2}{16bh^2} = \frac{3 \times 35 \times 100}{16 \times 0.6 \times 0.09 \times 10^3} = 12.2 \text{ N/mm}^2$$

Appendix B. The Effect of Local Buckling on the Collapse Load of a Fixed-ended Beam

The following analysis is introduced in Section 4.4.1 (p. 116), which explains its purpose and gives a summary and discussion of the results.

The fixed-ended uniform beam shown in Fig. B.1 is subjected to increasing load w per unit length. While elastic, it is assumed to have

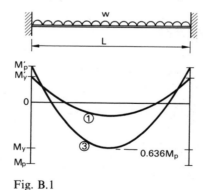

Fig. B.1

uniform flexural rigidity EI. The positive and negative moments of resistance are M_p and M'_p, and

$$M_p = 2M'_p$$

Positive and negative curvatures are ϕ and ϕ', and the moment–curvature relations for cross-sections, given in Fig. B.2, reproduce essential features of those found in composite beams. There is an initial elastic portion, OA. After first yield, here assumed to be at $0{\cdot}75M_p$, there is a fourfold increase in flexibility, giving the line AB. The 'plastic plateau' BC is followed by a falling branch CD, due at midspan to crushing of concrete and at the supports to buckling of the steel section.

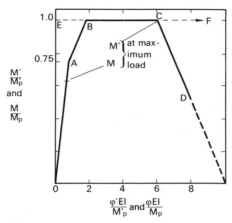

Fig. B.2

It is shown on p. 116 that simple plastic theory gives the collapse load w_p for this beam as

$$w_p = 12M_p/L^2 \qquad (4.14)$$

The calculation gives no information on strains or deflections prior to collapse, and assumes that all cross-sections follow the rigid–plastic moment–curvature relationship $OEBCF$ (Fig. B.2). It will now be shown that step-by-step calculation using elastic theory and the more realistic relationship $OABCD$ leads to the result that the collapse load is only $9{\cdot}1M_p/L^2$, and gives values of M and M' at all stages of loading (Fig. B.3).

Equilibrium requires that at all loads,

$$M + M' = wL^2/8 \qquad (B.1)$$

It can be shown from (4.14) that a useful consequence of this result is that in Fig. B.3 the vertical intercept between the dashed line OE and the line for M'/M_p' must always be twice that between OE and the line for M/M_p.

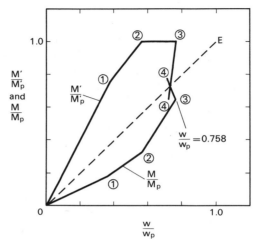

Fig. B.3

From the elastic analysis of a uniform fixed-ended beam, the end moment is initially twice the midspan moment, so yield first occurs at the supports, when the bending moments are as curve (1) in Fig. B.1. The load is given by

$$w_1 L^2/8 = (1 + 0.5) \times 0.75 M_p' \qquad \text{(B.2)}$$

so from (4.14),

$$w_1/w_p = 0.375 \qquad \text{(B.3)}$$

and M and M' are as given by points (1) on Fig. B.3. At this stage, M'/M_p' is 0.75 but M/M_p is only 0.187. This poor match between the elastic moments at hinge locations and the strengths available at these points (Fig. B.1) is a basic property of continuous composite beams, and explains why yield first occurs in this example at so low a proportion of the collapse load, and why the need for rotation capacity at hinge locations, to allow redistribution of moments, is greater than in steel beams of uniform section.

Redistribution causes the points of contraflexure to move along the beam; but for the present purpose it is accurate enough to assume that they are located at $0 \cdot 1 L$ from each support. The flexural rigidity of the end tenths of the span is now assumed to change from EI to $0 \cdot 25 EI$, so for the next increment of load the beam is as in Fig. B.4.

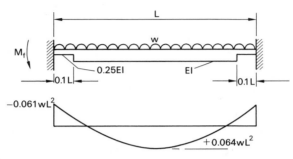

Fig. B.4

The moments given by elastic theory for such a beam are also shown. The available increase in end moment is $0 \cdot 25 M_p'$, giving

$$\delta w = 0 \cdot 25 M_p'/0 \cdot 061 L^2 = 4 \cdot 1 M_p'/L^2 \tag{B.4}$$

From (4.14) with $M_p = 2 M_p'$, $\delta w/w_p = 0 \cdot 171$, so the total load at this stage is

$$w_2/w_p = 0 \cdot 375 + 0 \cdot 171 = 0 \cdot 546 \tag{B.5}$$

The additional midspan moment is $0 \cdot 064 \times 4 \cdot 1 M_p'$, or $0 \cdot 131 M_p$, so at point (2) on Fig. B.3,

$$M/M_p = 0 \cdot 187 + 0 \cdot 131 = 0 \cdot 318 \tag{B.6}$$

Thus the negative-moment hinge forms at just over half the theoretical collapse load, when the midspan moment is less than one-third of M_p.

No further increase in M' is possible, so the next increment of loading is carried by increase in M alone, and is found to be limited by the rotation capacity of the negative-moment region. At stage (2) the curvature at each support is given by $\phi' EI/M_p' = 1 \cdot 75$ (point B on Fig. B.2), so the available increase (to point C) is $6 - 1 \cdot 75$, or $4 \cdot 25$, and $\delta \phi' = 4 \cdot 25 M_p'/EI$. We assume a linear variation of curvature increment from this value at each support to zero at the assumed points of contraflexure, so the available rotation in the negative moment region is

$$\theta = \frac{L}{20}\,\delta\phi' = \frac{L}{20} \times \frac{4 \cdot 25 M'_p}{EI} = 0 \cdot 106 M_p L / EI \qquad (B.7)$$

This occurs at constant moment M'_p, and so is assumed to be the end slope of the 'simply-supported' beam that carries the next load increment (Fig. B.5). From elastic theory, with $\delta M = \delta w L^2 / 8$,

$$\theta = \delta w L^3 / 24 EI = \delta M L / 3 EI \qquad (B.8)$$

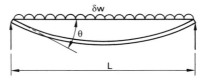

Fig. B.5

So from (B.7),

$$\delta M / M_p = 0 \cdot 318 \qquad (B.9)$$

and from (B.6), the total is $M/M_p = 0 \cdot 636$. The load increment is $8 \delta M / L^2$, so from (4.14) and (B.9),

$$\delta w / w_p = 8 \times 0 \cdot 318 / 12 = 0 \cdot 212$$

and at point (3) on Fig. B.3,

$$w_3 / w_p = 0 \cdot 758 \qquad (B.10)$$

Thus the negative-moment region enters the falling branch CD of its moment–curvature curve before yield occurs at midspan. The bending moments at this load are shown by curve (3) in Fig. B.1.

It is not now obvious whether further increase of deformation is associated with increase or decrease of total load; so the change of load associated with the increase in M/M_p from 0·636 to 0·75 (yield) is now calculated. Putting

$$\delta M = 0 \cdot 114 M_p \qquad (B.11)$$

then (B.8) gives

$$\theta = 0 \cdot 038 M_p L / EI \qquad (B.12)$$

From (B.7),

$$\delta\phi' = 20\theta/L = 0.76M_p/EI \tag{B.13}$$

From line CD in Fig. B.2,

$$\delta M'/\delta\phi' = -0.25EI$$

so from (B.13)

$$\delta M' = -0.19M_p = -0.38M_p' \tag{B.14}$$

From (4.14),

$$\delta w/w_p = 0.167\delta(M + M')/M_p$$

From (B.11) and (B.14),

$$\delta w/w_p = 0.167(0.114 - 0.19) = -0.013$$

so from (B.10)

$$w_4/w_p = 0.745$$

and point (4) can be plotted on Fig. B.3.

In this example, the beginning of the falling branch of the $M'-\phi'$ curve marked the maximum load, which, from (B.10), is only 76% of that given by simple plastic theory. Bearing in mind that the two lines (3)–(4) in Fig. B.3 must intersect on OA, it is obvious that even a substantial reduction in the slope of the falling branch CD in Fig. B.2 would make little difference to the failure load. An increase in the rotation capacity (BC) would be much more effective. It is also clear that the slope of the $M-\phi$ curve beyond point B is irrelevant.

References

1. Fisher, J. W. 'Design of composite beams with formed metal deck', *Eng. J., Amer. Inst. Steel Constr.*, **7**, 88–96, July 1970.

2. Johnson, R. P. and Cross, K. E. 'Design of low-cost composite structures for car parks', *Proc. Conf. Multi-storey and Underground Car Parks*, Institution of Structural Engineers, 81–87, May 1973.

3. *The Berkeley Hambro Bishopsgate Tower*, Project study 2, Constructional Steel Research and Development Organisation, London, Nov. 1973.

4. 'Fire protection', *Building with steel*, No. 3, 2–20, British Steel Corporation, May 1970.

5. *Modern Fire Protection for Structural Steelwork*, Publication FP 3, British Constructional Steelwork Association, London, 1967.

6. Kerensky, O. A. and Dallard, N. J. 'The four-level interchange between M4 and M5 motorways at Almondsbury', *Proc. I.C.E.*, **40**, 295–322, July 1968.

7. Johnson, R. P., Finlinson, J. C. H. and Heyman, J. 'A plastic composite design', *Proc. I.C.E.*, **32**, 198–209, Oct. 1965.

8. Cassell, A. C., Chapman, J. C. and Sparkes, S. R. 'Observed behaviour of a building of composite steel and concrete construction', *Proc. I.C.E.*, **33**, 637–658, April 1966.

9. CP 110, *The Structural Use of Concrete*, British Standards Institution, 1972.

10. BS 449: Part 2, *The Use of Structural Steel in Building*, British Standards Institution, 1969.

11. CP 117, *Composite Construction in Structural Steel and Concrete, Part 1: Simply-supported Beams in Building*, British Standards Institution, 1965.

12. CP 117, *Composite Construction in Structural Steel and Concrete, Part 2: Beams for Bridges*, British Standards Institution, 1967.

13. Rowe, R. E., Cranston, W. B. and Best, B. C. 'New concepts in the design of structural concrete', *Struct. Engr.*, **43**, 399–403, Dec. 1965.

14. Bate, S. C. C. 'Design philosophy and basic assumptions', *Concrete,* **7**, 43–44, Aug. 1973.

15. Draft code of practice: *Steel, Concrete, and Composite Bridges*, in preparation, British Standards Institution.

16. CP 3: Chapter 5; Part 1, *Dead and imposed loads*, 1967; Part 2, *Wind loads*, 1972, British Standards Institution.

17. Scott, W. B. *Composite Construction for Steel Framed Buildings—Properties of Composite Sections*, Publication 25, British Constructional Steelwork Association, London, 1965.

18. Ward, F. G. and Scott, W. B. *Composite Construction for Simply-supported Bridges—Properties of Composite Sections*, Publication BD 1, British Constructional Steelwork Association, London, 1967.

19. Johnson, R. P. 'Design of encased composite beams', *Consulting Engr.*, **32**, No. 11, 40–45, Nov. 1968.

20. Goble, G. G. 'Shear strength of thin-flange composite specimens', *Eng. J., Amer. Inst. Steel Constr.*, **5**, 62–65, April 1968.

21. Chapman, J. C. and Teraszkiewicz, J. S. 'Research on composite construction at Imperial College', *Proc. Conf. Steel Bridges*, 49–58, British Constructional Steelwork Association, 1969.

22. Johnson, R. P., Greenwood, R. D. and Van Dalen, K. 'Stud shear-connectors in hogging moment regions of composite beams', *Struct. Engr.*, **47**, 345–350, Sept. 1969.

23. Draft code of practice or British standard, *Composite Construction in Building*, in preparation, British Standards Institution.

24. Johnson, R. P. 'Research on steel–concrete composite beams', *Proc. A.S.C.E.*, **96**, ST3, 445–459, March 1970.

25. Yam, L. C. P. and Chapman, J. C. 'The inelastic behaviour of simply-supported composite beams of steel and concrete', *Proc. I.C.E.*, **41**, 651–684, Dec. 1968.

26. Johnson, R. P. 'Transverse reinforcement in composite beams for buildings', Synopsis, *Struct. Engr.*, **49**, 239, May 1971.

27. Johnson, R. P. 'Design of composite bridge beams for longitudinal shear', *Proc. Conf. Developments in Bridge Design and Construction*, Cardiff (March 1971), 387–399, Crosby Lockwood, 1971.

28. Mattock, A. H. and Hawkins, N. M. 'Shear transfer in reinforced concrete— recent research', *Journal Prestressed Concrete Institute*, **17**, No. 2, 55–75, March 1972.

29. Taylor, R., Plum, D. R. and Papasozomenos, A. G. 'Investigations on the use of deep haunches in composite construction', *Proc. I.C.E.*, **47**, 43–54, Sept. 1970.

30. Johnson, R. P. 'Design of composite beams with deep haunches', *Proc. I.C.E.*, **51**, 83–90, Jan. 1972.

31. Beeby, A. W. *The Prediction and Control of Flexural Cracking in Reinforced Concrete Members*, 55–75, Publication SP 30, American Concrete Institute, Detroit, March 1971.

32. Johnson, R. P. and Smith, D. G. E. 'Span–depth ratios for the control of deflections in composite beams', to be published, *Struct. Engr.*, **53**, Sept. 1975.

33. Johnson, R. P. and May, I. M. 'Partial-interaction design of composite beams', to be published, *Struct. Engr.*, **53**, August 1975.

34. Badoux, J.-C. *et al. Recommendations pour l'Utilisation de Tôles Profilées dans les Planchers Mixtes du Bâtiment*, Centre Suisse de la Construction Metallique, Zürich, 1973.

35. Johnson, R. P. and Climenhaga, J. J. 'Fatigue strength of form-reinforced composite slabs for bridge decks', *Publications, Int. Assoc. for Bridge and Struct. Engr.*, **35-I**, 89–101, 1975.

36. Aalami, B. and Williams, D. G. *Design of Thin Plates under Edge-compression*, Crosby Lockwood Staples, to be published.

37. PD 4064. *The Use of Cold Formed Steel Sections in Building*, Addendum No. 1 to BS 449:1959, British Standards Institution, 1961.

38. Resevsky, C. G. *Composite Slab Construction Theories and Practice*, M.B.S. Thesis, Department of Arch. Science, University of Sydney, 1970.

39. Bryl, S. 'The composite effect of profiled steel plate and concrete in deck slabs', *Acier-Stahl-Steel*, No. 10, 448–454, Oct. 1967.

40. Allen, D. L. 'Vibrational behaviour of long-span floor slabs', presented at Canadian Structural Engineering Conference, Toronto, Jan. 1974. Canadian Steel Industries Construction Council, Willowdale, Ontario, 1974.

41. BS 153, *Steel Girder Bridges*, British Standards Institution, 1958.

42. CP 114, *Reinforced Concrete in Buildings*, British Standards Institution, 1965.

43. Moffat, K. R. and Dowling, P. J. *Parametric Study of the Shear Lag Phenomenon in Steel Box Girder Bridges*, CESLIC Report BG 17, Imperial College, London, Sept. 1972.

44. Baker, J. F. and Heyman, J. *Plastic Design of Frames*, 2 vols., C.U.P., 1969.

45. Horne, M. R. *Plastic Theory of Structures*, Nelson, 1971.

46. Johnson, R. P. and Willmington, R. T. 'Vertical shear strength of compact composite beams', *Proc. I.C.E.*, Supp. Vol., 1–16, Jan. 1972.

47. Johnson, R. P. and Willmington, R. T. 'Vertical shear in continuous composite beams', *Proc. I.C.E.*, **53**, 189–205, Sept. 1972.

48. *Joint Committee on Fully-Rigid Multi-Storey Welded Steel Frames*, Second Report, Instn. of Struct. Engrs., May 1971.

49. Yam, L. C. P. and Chapman, J. C. 'The inelastic behaviour of continuous composite beams of steel and concrete', *Proc. I.C.E.*, **53**, 487–502, Dec. 1972.

50. Barnard, P. R. and Johnson, R. P. 'Plastic behaviour of continuous composite beams', *Proc. I.C.E.*, **32**, 180–197, Oct. 1965.

51. Climenhaga, J. J. and Johnson, R. P. 'Moment–rotation curves for locally buckling beams', *Proc. A.S.C.E.*, **98**, ST6, 1239–1254, June 1972.

52. Climenhaga, J. J. and Johnson, R. P. 'Local buckling in continuous composite beams', *Struct. Engr.*, **50**, 367–374, Sept. 1972.

53. BS 4360, *Weldable Structural Steels*, British Standards Institution, 1972.

54. Hope Gill, M. *Ultimate Strength of Continuous Composite Beams*, Ph.D. Thesis, University of Cambridge, 1974.

55. Comité Européen du Beton and Fédération Internationale de la Précontrainte, *International Recommendations for the Design and Construction of Concrete Structures*, Cement and Concrete Association, London, June 1970.

56. Johnson, R. P. and Buckby, R. J. 'Control of cracking', Technical Paper to Sub-committee B/116/5, British Standards Institution, May 1974.

57. Faber, O. 'More rational design of cased stanchions', *Struct. Engr.*, **34**, 88–109, March 1956.

58. Stevens, R. F. 'Encased stanchions and BS 449', *Engineering*, **188**, 376, Oct. 1959.

59. McDevitt, C. F. and Viest, I. M. (a) 'Interaction of different materials', Introd. Report, 55–79. (b) 'A survey of using steel in combination with other materials', Final Report, 101–117, Ninth Congress, Int. Assoc. for Bridge and Struct. Eng., Amsterdam, 1972.

60. ACI 318:1971, *Building Code Requirements for Reinforced Concrete*, American Concrete Institute, 1971.

61. Jones, R. and Rizk, A. A. 'An investigation on the behaviour of encased steel columns under load', *Struct. Engr.*, **41**, 21–33, Jan. 1963.

62. Stevens, R. F. 'Encased stanchions', *Struct. Engr.*, **43**, 59–66, Feb. 1965.

63. Basu, A. K. and Sommerville, W. 'Derivation of formulae for the design of rectangular composite columns', *Proc. I.C.E.*, Supp. vol., 233–280, 1969.

64. Virdi, K. S. and Dowling, P. J. 'The ultimate strength of composite columns in biaxial bending', *Proc. I.C.E.*, Part 2, **55**, 251–272, March 1973.

65. *Design Manual for Concrete-filled Hollow Section Steel Columns*, Cidect Monograph No. 1, British Steel Corporation, 1970.

66. Johnson R. P. and Anderson, D. 'Design studies for composite columns', Unpublished Report, University of Warwick, Jan. 1974.

67. Bondale, D. S. and Clark, P. J. 'Composite construction in the Almondsbury interchange', *Proc. Conf. Structural Steelwork*, 91–100, Sept. 1966. British Constructional Steelwork Association, 1967

68. Neogi, P. K., Sen, H. K. and Chapman, J. C. 'Concrete-filled tubular steel columns under eccentric loading', *Struct. Eng.*, **47**, 187–195, May 1969.

69. Furlong, R. W. 'Design of steel-encased concrete beam-columns,' *Proc. A.S.C.E.*, **94**, ST1, 267–281, Jan. 1968.

70. Sen, H. K. *Triaxial Effects in Concrete-filled Tubular Steel Columns*, Ph.D. Thesis, University of London, July 1969.

71. Johnson, R. P. and Hope-Gill, M. 'Semi-rigid joints in composite frames', Prelim. Report, Ninth Congress, Int. Assoc. for Bridge and Struct. Eng., 133–144, Amsterdam, May 1972.

72. Wood, R. H. *A New Approach to Column Design, with Special Reference to Restrained Steel Stanchions*, Note 263/71, Building Research Station, 1971; published, HMSO, 1974.

Index